Reprint Publishing

FOR PEOPLE WHO GO FOR ORIGINALS.

www.reprintpublishing.com

ON MILDEW

AND

FERMENTATION.

BY

A. DE BARY.

BERLIN 1872.

C. G. LÜDERITZ'sche Verlagsbuchhandlung.

CARL HABEL.

33 Wilhelm Strasse 33.

LONDON,

WILLIAMS & NORGATE

14 Henrietta Street, Covent Garden.

NEW-YORK,

B. WESTERMANN & Co.

GERMAN AND FOREIGN BOOKSELLERS,

524 Broadway.

In the economy of the whole animal creation, the principal function of the vegetable kingdom is to produce from the constituent parts of water, air and decomposed minerals the organic combination of carbon with hydrogen, oxygen and nitrogen, from which the bodies of living creatures, and even of plants and animals are constructed.

Vegetation alone produces this matter; animals consume it and change it, at last, again into the simple unorganized combination of carbonic acid, water, ammoniac etc. etc., which formed the first material for their construction. The same result follows the process of putrefaction and decay, into which vegetable bodies also fall after they are dead.

This organic creation of constructive activity belongs especially to green coloured vegetation; it is a function of the *chlorophyll*, the green matter of the leaves of vegetables and plants. Even on the green leaved plants, the parts of which are not green, are only spots in which the organic matter is changing or being deposited, not its original place of formation.

Generally, when vegetation is spoken of, the green coloured is meant; we forget thereby, that this although a very powerful, is, at the same time, only a fraction of the whole vegetable kingdom. A great number of plants are always entirely destitute of chlorophyll or an equivalent colour, and are in consequence of this deprived of the power to produce organic matter from unorganic raw material. Therefore, like to the animals, it is allowed to them to make use of already existing organic substances for their existence, which is derived directly or indirectly from green vegetation. They do not thrive in that soil which proceeds from the disintegration of the stones, and is simply saturated with water and surrounded by air, which is sufficient for the green vegetable kingdom; but they require much more an already prepared fostering soil, made from the organic bodies of plants and animals. They settle themselves thus either on the living organism itself as parasites, or on their withered parts and products, which have fallen into decay, as putrified plants, Saprophytes.

The most diverse classes and divisions of the vegetable kingdom, furnish their contingent to the vegetation which is free from chlorophyll. In tropical climates, floriferous parasitical plants of many various forms are found, which have no green colour, and in our country the Orobache, which grows on hemp, tobacco and trefoil, and the dodder (Cusacta) are examples which are universally known. There are also specimens of *Saprophytes* which have no chlorophyll, among the most various kinds of floriferous plants, such as heaths, gentian, orchidiæ etc. Of our native plants, we can bring forward the Bird's Nest (Neottia nidus avis), and the yellow birdsnest (Monotropa), a parasitical plant growing on the decayed stumps of trees in fir woods, as the well-known inhabitants of mouldering wood-grounds.

By far the greater number of plants which are free from chlorophyll, belong to the non-floriferous cryptogamic species and these are generally designated altogether by the name of *Fungi*, although it is not quite correct to do so. We will for the present concede to the popular application, in order to rectify it later.

It is not necessary to assure any one who has once beheld the ground of a forest, in a damp autumnal season, that the Fungi exhibit an immense variety of forms and species, and yet the large quantities which present themselves to the cursory glance, are only a few of the whole number, as the greater part of the Fungi are so microscopically small, that they are very difficult to discern, or are indeed scarcely visible to the naked eye. In the present state of our knowledge, it is not possible to say exactly, how many species of Fungi there are, or how many are known, but when we consider, that very many kinds of floriferous plants have at least one Fungus as a parasite or saprophyte, and if we reckon thereto the species which appear elsewhere as well as on floriferous plants, it is no exaggerated estimation, if we place the number of the species of living Fungi on an equality with that of the floriferous plants; viz: about 150,000. Every attentive examination further proves, that the greater part of the species of Fungi are not inferior in number to the floriferous plants. We can therefore, without fear of error, consider, that the vegetation of which we have just spoken, is, at least, equally rich and various as that of the florifarous and chlorophyll, although it is greatly inferior to them in number.

Fungi are to be found everywhere; their roots are established on, between and near those of other plants; there is scarcely a spot in which organic bodies are found, that does not serve as an abode for Fungi.

It is to be supposed, as a matter of course, that such a

rich and widely spread group of organism must in some way powerfully act on the economy of Nature, and on closer observation we find, that its function is that of a police-man.

The parasitical Fungi attack chiefly single species of plants and animals, which are generally of that kind which is suitable for their nourishment. They transplant themselves by means of their shoots to new detached species, as long as these live scattered among kinds which are indifferent to the parasite. Those plants which are attacked by the parasite, become sickly and their decay is accelerated. The more a species multiplies, which nourishes a parasite, the more exclusively and densely it takes possession of a space, at the cost of others, the easier is the parasite and the disease which is caused by it, conveyed from one plant to the other, and the disease takes the character of an epidemic. The epidemical illnesses of many plants, which are found exclusively in cultivated pieces of ground, and also of very many common ones which grow wild, afford us well-known examples of this fact; innumerable caterpillars, flies etc. are killed every year through the parasitical Fungi. The police service of the parasites is directed against the increase of some sociable species at the cost of others.

This zeal, however, has little weight against the energetic activity of the scavengers, which is directed towards the eradicating of all the dead organic substance of vegetating saprophytes.

If the decayed organic bodies are exposed to the atmospheric air of a certain temperature, and are in the neighbourhood of water, disunions occur, which render the combined complicated connection, which they possessed in life, much more simple. The result is putrefaction: the organic substance disappears, because it is burnt to carbonic acid, water and ammoniac, and escapes in the surrounding air, while the relatively small portion of incombustible (mineral) ingredients remains behind as ashes. This putrefaction is generally brought about by a division into new combinations which are more simple than the original organization, but different from those of the last products of decay; this is called putrefaction and fermentation; names which in no way answer to any distinct idea, but merely express a conventional difference in every day life, according to the material which is decomposed, and the qualities of certain decomposing products; thus we say, the *must ferments,* and *stale meat putrifies.*

The process of fermentation and putrefaction can be produced in laboratories by many various methods. They can also be brought about in Nature in manifold kinds. But if we observe minutely, we perceive (as we will slightly notice for the present, but later more particularly discuss), that the preponderating number of these processes are actually excited and maintained by the process of vegetation in the Fungi universally found in decomposing materials. Without this activity of the Fungi, the dead animal and vegetable bodies would undergo a slow oxydation, collect together in masses on the surface of the earth, and very soon hinder the existence of life, instead of quickly making room for new generations, and at the same time restoring to the circulation carbonic acid, water and ammoniac, the nutritious matter of every life.

Every species of Fungus has like all other animal and vegetable kinds its peculiar conditions of vegetation; according to which the different species frequently share among themselves the various sorts of decomposing material; still very often several different kinds dwell in the same soil. Corresponding to these diversities, many species excite in the same substrate the same or similar decompositions; while partly to particular kinds belong particular decomposing results. Those which, according to their form, are of different kinds, can, in complience with this view, act differently or similarly.

The physiological characteristics, which we have hitherto concisely mentioned, the variety and that which may be added here, viz: their frequently occuring wonderful growth and development of form, imbue the Fungi with a very high scientific interest. To many is also added their practical use in the economy of life. We know, that many of our cultivated plants become deseased and are destroyed by parasitical Fungi, and that they threaten with destruction many useful animals, for example, the silk-worm, and even the human body. A number of fermenting, putrefying and decaying processes play an important part in the economy of human life, sometimes threatening danger and destruction, sometimes proving very serviceable, and these effects are generally produced by Fungi. These phenomena have been more clearly explained, and may advantages have been attained from seeking and carefully studying the Fungi which cause them. It is no wonder, therefore, that, wherever disease, putrefaction etc. make their appearance, people seek for Fungi, and in their blind zeal often declare the first thing they find, which resembles a Fungus, to be the evil-doer, which had occasioned the long list of phenomena, which until then had remained totally incomprehensible. This modern hunt for Fungi will induce every one who has an interest in Natural Philosophy, to ask what is actually known of the life and effects of the Fungi. The attempt to answer this question, if only in part, shall be made in the following pages.

In order to be able to judge of the effects which a living organism exercises through its animal economy, it is above all necessary to understand this organism itself, with its form and method of development. Great importance will be laid on an accurate description of those parts. It is often a dry and wearisome subject, because many of the Fungi have a very complicated course of development, and because, very often. the same species appears under different forms, which produce one another by turns, or displays propagating organs of different constructions, which are formed by the same plant in more or less regular succession and, in consequence of this, controversies can arise about forms which grow together, whether they belong to the sphere of development of the same or different species; — controversies the discussion of which is often unavoidable, if we wish to obtain a clear view of many questions of general interest. The time allotted to the lectures is by far too limited, to allow us to discuss all these subjects with that minuteness, which is indispensable to ex- plicitness, or to amplify the several topics of discourse, or even to discribe the principal representatives of all the groups of Fungi. It is much more desirable to limit our attention to a small number of examples, and as such I have chosen those which are to be found in dead organic bodies, and are generally called *Mildew* and *Fermentation.*

Mildew and fermentation are not scientific expressions, but a more accurate definition of these names is not to be found. If we would give something, which at a distance may appear like a definition, we must only follow the practise of daily life, and designate by the name mildew that fleecy, threadlike fungus formation, which appears in organic bodies which are in a state of decay; and by the name fermentation, that substance which is found on the before mentioned bodies as a cloudy or greasy deposit or crust. The former consists of long thread-like ramous pipes or similarly shaped rows of *closely* connected cells, the latter mostly, but not always, of accumulated little cells, *loosely* connected together.

But even among the more particularly indicated objects, we must confine the following examination within still greater limits. The same reasons which were asserted above, for the concentration of our description into a relatively small sphere, forbid us here to search into the whole series of well-known forms, which are generally called *Mildew* and *Fermentation*. We can only take a few examples from them, and will choose for reasons which may easily be understood, those which are the most widely spread and abundant, and which have the greatest affinity to our daily life and, at the same time, the so-called *fungi question*.

The course of the lecture will justify us in commencing with the form of *Mildew*.

In every household, there is a frequent unbidden guest, which appears particularly on preserved fruits, viz: the *Mildew* or *mould* fungus, which is called *Aspergillus glaucus* or also *Eurotium herbariorum* (Fig. 1). It shows itself to the naked eye as a woolly, flocky crust over the substrate, first purely white, then gradually covered over with little fine glaucous or dark green dusty heads. More minute microscopical examinations show, that the fungus consists of richly ramified fine filaments, which are partly disseminated in the substrate, and partly raised obliquely over it. They have a cylindric form with rounded ends, and are divided into long outstretched members, each of which possesses the property which legitimizes it as a *vesicle* in the ordinary sense of the word; it contains enclosed within a delicate structureless wall those bodies which bear the appearance of a finely granulated mucous substance, which is designated by the name *Proto-plasma*, and which either equally fills the cells, or the older the cell, the more it is filled with watery cavities called *Va-cuolæ*.

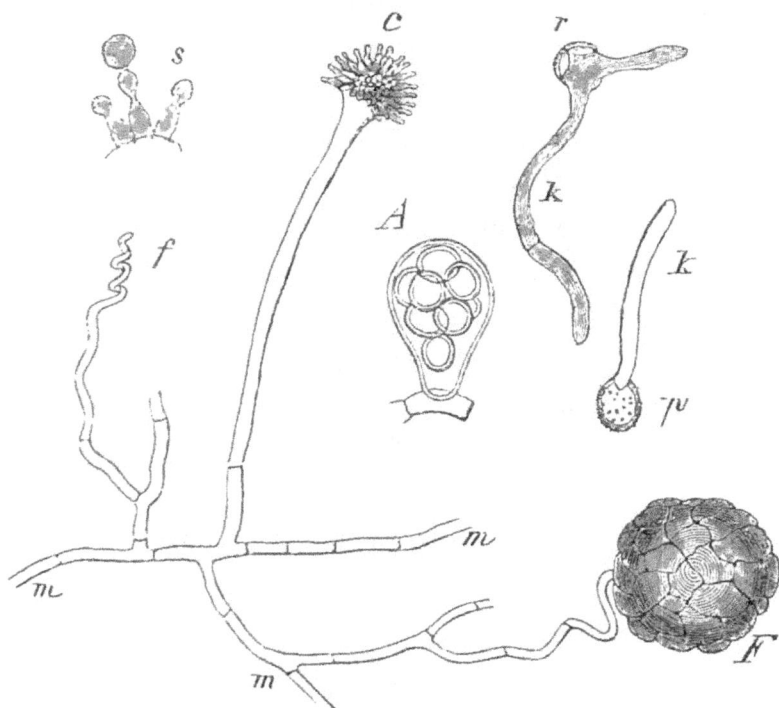

Fig. 1. *Aspergillus glaucus.*

mm Mycelium thread or filament, a conidium stamen (c) from which the conidia have fallen away. F an Utricle in perfection, magnified 190 times. f the beginning of an utricle. s three sterigma from the summit of a conidia stamen, showing the unlacing of the spores. p Germinating conidii (enlarged 250—300). A the ampula for the spores (enlarged 600 times). r germinating ampula. k amnios.

All parts are at first colourless. The increase in the length of the filaments takes place through the preponderating growth near their points; these continually push forward, and at a short distance from them, successive new partitions rise up, but at a greater distance, the growth in the length cease. This kind of growth is called "point growth". The twigs and branches spring up as lateral dilatations of the principal filament, which once designed, enlarges according to the "point growth", which has been already described. The "pointgrowth" of every twig is, to a certain extent, unlimited.

The filaments in and on the substrate (m, k), which have been already described, are the first existing members of the fungus, they continue so long as it vegetates, as the parts which absorb nourishment from and consume the substrate; they are called *Mycelia*. Nearly every fungus possesses a *mycelium*, which, without regard to the specific difference of form and size, especially shows the before mentioned nature in its construction and growth. To avoid all unnecessary repetitions, we have so fully described it here, once for all.

The superficial threads of the Mycelium of our fungi, produce other filaments besides those numerous branches which have been already described, which are fruit stamens, or more specially designated, conidia stamens (c). These are on an average thicker than the Mycelium threads, and only exceqtionally ramified or furnished with partitions; they rise almost perpendicularly in the air, and attain a length of, on an average, half a millimeter; but they seldom become longer, and then their growth is at an end. Their free upper end swells to the form of a round *spathe*, and this produces on the whole of its upper part rayed diverging protuberances, which attain an oval form and a length, which almost equals their radius or in weaker specimens, the diameter of the round spathe. The rayed diverging protuberances are the direct producers and bearers of the propagating cells, spores or conidia, and are called *Sterigmata*. Every sterigma at first produces at its point, a little round protuberance, which with a strong narrowed basis, rests upon the sterigma. These are filled with protoplasma, swell more and more and, after some time, separate themselves by a partition from the sterigma into independent cells — spores or conidia. The formation of the first spore takes place on the same end of the sterigmen, and in the same manner a second follows, then a third, and so on; every one, which springs up later, pushes its predecessor in the direction of the axis of the sterigma in the same degree in which it grows itself; every successive spore formed from a sterigma, remains for a time in a row with one another. Consequently every sterigma bears on its apex a chain of spores, which are so much the older, the farther they stand from the sterigma. The number of the links in a chain of spores reaches in normal specimens to ten or more. All sterigmata spring up at the same time, and keep pace with one another in the formation of the spores. The round apex of the stamen is at last covered with a thick head of radiated chains of spores. Every spore grows for a time, according to its construction, and at last separates itself from its neighbours. The mass of dismembered spores forms that fine glaucous which we have mentioned above. The process, which has already been described, of the formation of spores, by which a protuberance of the sterigma dismembers itself to an independent spores-cell, and at last separates itself entirely, is called, according to its appearance, *articulation*;

in the case with which we are now occupied, the spores are
articulated in rows one after the other from the ends of the
sterigmata. This expression will often be used further on. The
ripe conidion is a cell of a round or broad oval form, generally about $\frac{1}{95}$ Mm. large, filled with a colourless protoplasma
and, if observed separately, is found to be provided with a
brownish fine verruculose dotted wall (p).

The same Mycelium, which forms the pedicle for the
conidia, when it is near the end of its development, forms by
normal vegetation a second kind of fructification, which is to
be described as the bearer of the Utricle. It begins as delicate thin little twigs, which are not to be distinguished by
the naked eye, and which, mostly in 4 or 6 turns, after a
quickly terminated growth, wind their ends like a cork-screw.
The sinuations decrease in width more and more, till they at
last reach close to one another, and the whole end changes
from the form of a cork-screw into that of a hollow screw.
In and on that screw-like body, changes of a complicated
kind take place, the detailed description of which would here
occupy too much time, and therefore we will only mention,
that they must be designated as a productive process. In
consequence of the same, from the screwed body, a globular
receptacle (Utricle, F) is formed, consisting of a thin wall of
delicate cells, and a closely entwined row of cells surrounded
by this dense mass. By the enlargement of all these parts,
the round body grows so much, that by the time it is ripe,
it is visible to the naked eye. The outer surface of the wall
assumes a compactness and a bright yellow colour. The
greater part of the cells of the inner mass (a quantity are
dissolved in favour of the rest) become ampullæ for the formation of spores (*spores ampullæ*, *Asci*), while they free

themselves from the reciprocal union, take a broad oval form and each one produces in its inner space eight spores (A). These soon entirely fill the space of the ampulla. When the spores are quite ripe, the ampulla disappears, the wall of the utricle becomes brittle, and from the irregular rents, which easily arise from contact, the colourless round spores are liberated.

With the ripeness of the Utricle, the Mycelium generally takes a yellow or yellow-red colour; therefore the whole fungi covering has a different appearence; the growth of the Mycelium ceases, at the same time, other distinct characteristic products of development, than those already described, do not show themselves, but very often innumerable stunted and curiously formed conidia pedicles, which are very easy to recognize.

Till now, the term Spore has been used without any explanation of its meaning. It denotes here and elsewhere the cells, which separate themselves from the mother-plant and become new individuals, not the direct products of a sexual procreation. The Utricles are, as we have already intimated, products of a sexual production, and also the Asci, which are formed from them, but the spores are not generated by impregnation; therefore, we apply to them the name which we have used in the foregoing definition. We also find two sorts of spores in one species, that which is formed in the ampulla and that which is separated from the sterigma in rows; it is necessary here as well as in other cases, to distinguish both kinds from one another, by particular names, and it is now agreed on to designate those which are detached from free thread-like pedicles, with the name of *Conidia,* and the others *Utricle spores* or *Asco spores.*

The pedicles of both kinds of spores are formed from the same Mycelium in the order already described. If we examine very attentively, we can often see both springing up close to one another from the same filament of a Mycelium. This is not very easy in the close interlacing of the stalks of a mass of fungi, in consequence of their delicacy and fragility. Before their connection was known, the Utricles and conidia pedicles were considered as organs of two very different species of fungi, and the Utricles were called *Eurotium* and the others *Aspergillus*; this is the origin of the double name.

The perfect development of form of a fungus like that of every other organism, naturally depends on particular external conditions. If these be only partly to be found, then the development is imperfect. For this reason, we often find our Aspergillus only a Conidion, bearing no utricle; the latter always fails, if it be intentionally badly nourished. The reverse case, that the Mycelium produces only utricles and no conidia, is not known, and certainly never takes place.

It remains at last to legitimize the spores of fungi as cells which serve for propagation.

A detailed description of all their peculiarities and at the same time, the particular form of the utricle spores, can be here omitted, and a reference to that which we have above mentioned, will be sufficient.

If we sow both spores on a suitable substrate such as a solution of sugar, the juice of fruit or the damp surface of bodies already taken possession of by fungi in a perfect state, they begin to vegetate, that is, they perceptibly swell and drive out, like most of the fungi spores, on one or two sides a pipe-like cylindrical protuberance, called a germinating spore, into which the protoplasma of the spore gradually finds its way and which, presuming the nourishment is sufficient, immediately grows up to a mycelium filament possessing the above described properties (p, r). These first form conidia pedicles, and later utricles. The products of both kinds of spores are in all essential respects exactly alike.

With respect to the flourishing of the *Aspergillus glaucus*, it is to be found on the dead parts of plants, mostly as it has already been mentioned, at the commencement of its development, and particularly, when it is very luxuriant; also on other dead organic bodies of manifold kinds. It is sometimes to be found in diseased organs of living animals and human beings, particularly in the acoustic ducts of diseased ears, in the air passages of birds etc. It is not known as a parasite which attacks living bodies and diseases them. In the latter named places it is also only found as conidion, never with utricles. In the same places, in which A. glaucus is found, exist many more similar forms, of which several species are distinguished by the names *Asp. niger*, *Asp. repens* etc.

We will now describe the *Botrytis cinerea* (Fig. 2). We shall see, that this fungus is very widely spread on many different substrates; we will, however, for reasons which shall be explained later, entirely confine ourselves, for the present, to one, on which it is almost always to be found, viz: on dead, damp fallen vine-leaves.

It spreads its Mycelium in the tissue, which is becoming brown, and this shows at first (without regard to the specific peculiarities, which we have here justly neglected), essentially the same construction and growth, as that which we have described for the Mycelium filaments of the Aspergillus.

On the Mycelium soon appear, besides those which are

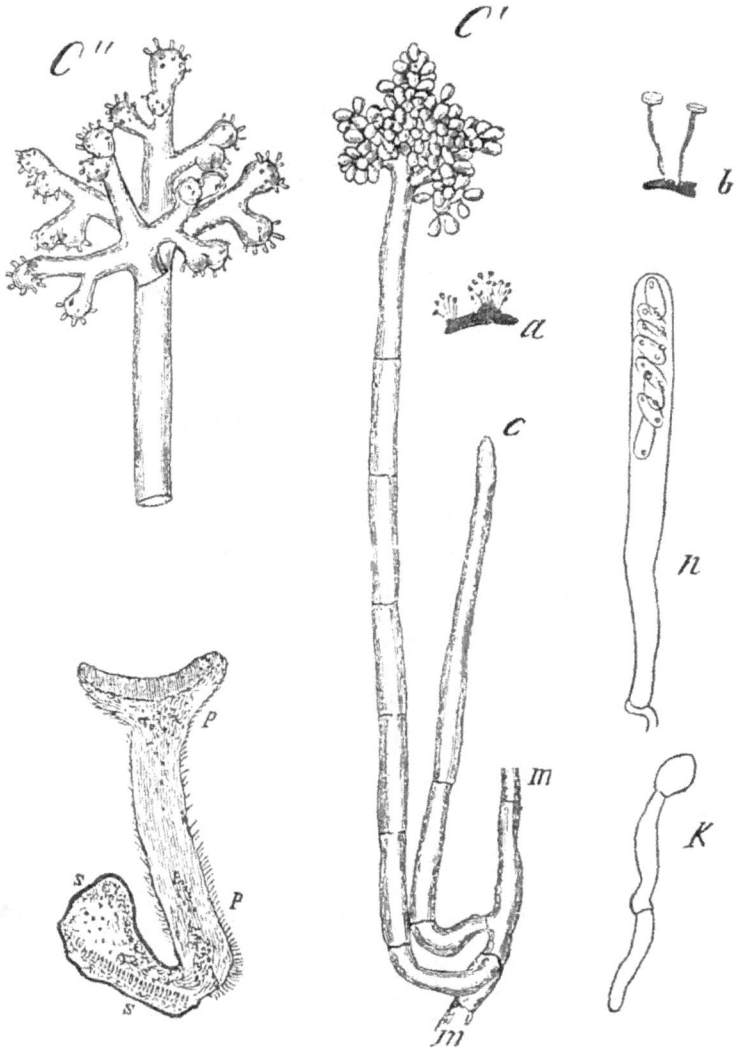

Fig. 2. *Botrytis cinerea.*

a and b natural size; Sclerotia, from which grow by a conidia bearers, by b utricles.— c c′ conidia bearers (c′ with ripe conidia) springing from the Mycelium filament m (magnified about 200 times). — C″ the end of a conidion bearer with the first beginning of conidia articulation on the branches.— k germinating conidia (magnified 300 times).— d (slightly magnified) section of a sclerotium s, from which grows a very little utricle (p, p).— u (magnified 390 times) single spore with 8 ripe spores.

spread over the tissue of the leaves, strong thick, mostly fas-
cicular branches, which stand close to one another, breaking
forth from the leaf and rising up perpendicularly; the Conidia
bearers (C C′). They grow about 1 Mm. long, divide themselves by successively rising partitions into some procumbent
cylindrical linked cells, and then their growth is ended and
the upper cell produces near its point 3—6 branches almost
standing rectangular. Of these the under ones are the longest
and they again shoot forth from under their ends one or
more still shorter little branches. The nearer they are to the
top, the shorter are the branches and less divided; the upper
ones are quite branchless, and their length scarcely exceeds
the breadth of the principal stem. Thus a system of branches
appears upon which, on a small scale, a peniculate efflores-
cence is found and geniculated somewhat in the form of a
grape (C″). All the twigs soon end their growth; they all
separate their inner space from the principal stem, by means
if a cross-partition placed close to it. All the ends and also
that of the principal stem swell about the same time; something like a bladder and on the upper free half of each
swelling appear again simultaneously several (about 6—10)
fine protuberances close together, which quickly grow to little
oval bladders filled with protoplasma and resting on their
bearers with a subsessile pedicellate narrow basis, and which
at length separate themselves through a partition according to
the manner described in the process of the Aspergillus. The
detached cells are the *Conidia* of our fungus; only one is
formed on each stalk. When the formation is completed in
the whole of the panicle, the little branches which compose
it, are deprived of their protoplasma in favour of the con-
sumed Conidia; it is the same with the under end of the

principal stamen, the limits of which are marked by a cross-partition. The delicate wall of these parts shrinks up until it is unrecognizable; all the Conidia of the panicle approach one another to form an irregular grape-like bunch which rests loosely on the bearer C', and from which it easily falls away as dust. If they be brought into water, they fall off immediately; only the empty shrivelled delicate skins are to be found on the branch, which bore them, and the places, on which they are fixed to the principal stem, clearly appear as round circumscribed *hilums*, generally rather arched towards the exterior.

The development of the main stem is not ended here. It remains solid and filled with protoplasma as far as the portion which forms the end through its Conidia. Its end which is to be found among these pieces, becomes pointed after the ripening of the first panicle, pushes the end of the shrivelled member on one side, and grows to the same length as the height of 1—2 panicles, and then remains still, to form a second panicle similar to the first. This is later equally perfoliated as the first, then a third follows, and thus a large number of panicles are produced after and over one another on the same stamen. In perfect specimens, every perfoliated panicle hangs loosely to its original place on the surface of the stamen, until by shaking or the access of water to it, it falls immediately into the single conidia, or the remains of branches, and the already mentioned oval hilums are left behind. Naturally, the stamen becomes longer by every perfoliation; in luxuriant specimens the length can reach that of some lines. Its partition is already, by the ripening of the first panicle, from the beginning of its foundation, strong and brown; it is only colourless at the end which is extending, and in all new formations. During all these changes, the filament remains either unbranched, except as regards the transient panicles, or it sends out here and there at the perfoliated spots, especially from the lower ones, one or two strong branches, standing opposite one another and resembling the principal stem.

The Mycelium, which grows so exuberantly in the leaf, often brings forth many other productions, which are called *Sclerotia* and are, according to their nature, a thick bulbous tissue of Mycelium filaments. Their formation begins with the profuse ramification of the Mycelium threads in some place or other, generally but not always in the veins of the leaf; the intertwining twigs form an uninterrupted cavity, in which is often enclosed the shrivelling tissue of the leaf. The whole body swells to a greater thickness, than that of the leaf, and protrudes on the surface like a thickened spot. Its form varies from circular to fusiform; its size is also very unequal, ranging between a few lines and about half a Mm. in its largest diameter (a, b). At first it is colourless, but afterwards its outer layers of cells become round, of a brown or black colour, and it is surrounded by a black rind consisting of round cells, which separate it from the neighbouring tissue. The tissue within the rind remains colourless; it is an entangled uninterrupted tissue of fungus filaments, which gradually obtain very solid hard cartilaginous coats. The sclerotium, which ripens, as the rind becomes black, loosens itself easily from the place of its formation, and remains preserved, after the latter is decayed.

The sclerotia are — here as in many other fungi — perennial organs, designed to begin a new vegetation after a state of apparent quietude, and to send forth special fruit-bearers. They may in this respect be compared to the bulbs and perennial roots of under shrubs.

The usual time for the development of the Sclerotia is late in the autumn after the fall of the vine leaves. As long as the frost does not set in, new ones continually spring up, and each one attains to ripeness in a few days. If frost appears, it can lie dry a whole year, without losing its power of development. This latter commences, when the Sclerotium is brought in contact with damp ground during the usual temperature of our warmer seasons. If this occur soon, at the latest some weeks after it is ripe, new vegetation grows very quickly, generally after a few days: in several parts the colourless filaments of the inner tissue begin to send out clusters of strong branches, which breaking through the black rind, stretch themselves up perpendicularly towards the surface, separate from one another, and then take all the characteristics of the conidia-bearers (a). Many such clusters can be produced on one Sclerotium, so that soon the greater part of the surface is covered by filamentous conidia-bearers with their panicles. The colourless tissue of the Sclerotium disappears in the same degree as the conidia-bearers grow, and at last the black rind remains behind empty and shrivelled. If we bring after many months for the first time the ripe sclerotia in damp ground in summer or autumn, after it has ripened, the further development takes place more slowly as in the first mentioned case, and in an essentially different form. It is true, that from the inner tissue numerous filamentous branches shoot forth at the cost of this growing fascicle, and break through the black rind, but its filaments remain strongly bound, in an almost parallel situation, to a cylindrical cord, which for a time lengthens itself and spreads out its free end to a flat plate-like disk. This is always formed of strongly united threads, ramifications of the cylindrical cord (d). On the free upper surface of the disk, the filaments shoot forth innumerable branches, which growing to the same height, thick and parallel with one another, cover the before named disk. Some remain narrow and cylindrical, are very numerous and produce fine hairs (Parophyses); others also very numerous, take the form of club-like ampulla cells, and each one forms in its interior eight free swimming oval spores (n). Those ampulla cells are sporuliferous *Asci* in the same sense of the word as that described for the *Aspergillus*, the stalked disk of the utricle of our fungus. After the spores have become ripe, the free point of the utricle bursts and the spores are scattered to a great distance by a mechanism which we will not here further describe. New ampullas push themselves between those which are ripening and withering, a disk can under favourable circumstances always form new spores for weeks at a time. The number of the already described utricle bearers is different, according to the size of the Sclerotium. Smaller specimens usually produce only one, larger 2—4 (b). The size is regulated by that of the Sclerotia and ranges in full grown specimens between one and more Mm. for the length of the stalk, and $\frac{1}{2}$—3 (seldom more) Mm. for the breadth of the disk.

We must at last state that, which is here of special interest with regard to the further development of the ripe conidia and utricles, that both in suitable media (in clear water it is impossible or very difficult) for example on the damp and injured surface of vine leaves, shoot forth germinating utricles, just like those of the *Aspergillus* (k), and the utricles grow directly into Mycelium filaments, which can form Conidia bearers and at last again become Sclerotia.

It may appear extraordinary, that an especial inhabitant of grape leaves should be described as an example of an universally spread mildew fungus. The Mycelium and Conidia bearers of the *Botrytis cinerea* or more properly said, such as cannot be distinguished from those already described, are indeed universally spread Mildew on the dead parts of plants of every kind — putrifying grapes, "musty" plants in damp hot-houses are places, in which it scarcely ever fails, ripe pumpkins, the dead stalks of the most different plants filled with sap, are often covered with it in the space of several square-inches. Not on the last named substrates do we miss Sclerotia which are often somewhat larger than those just described, but in every other respect have the same form. We also see, that Conidia bearers are produced in masses from these Sclerotia, in the same manner as that which we have already described. Utricle bearers with stalks have, on the contrary, been seldom observed, and when they are found, they are always very similar to those which originate in vine leaves, although in many instances very different from them. It is possible, that these various utricle forms belong to different, although very nearly related species; these species have shown till now no decided difference in the Conidia bearers and the Sclerotia, or in other words, the universally spread Conidia bearers which we must now all assign to the *Botrytis cinerea*, belong to some nearly related species of fungus, which are distinguished by their utricles.

This is the reason, why the description given above, applies more particularly to the one form of the grape, well known in its course of development. We can now add to that which we have already said, that from several vegetable substrates, only such attain the Sclerotia formation, which are in some measure of a massive and solid texture like many leaves, pumpkins, thick stalks etc. On very delicate and deciduous parts, as for example flowers, the sclerotium formation does not exist; only Mycelium threads and generally very luxuriant Conidia bearers are formed; therefore the fungus is only produced on these substrates in the one form which we have just mentioned.

It has been for a long time with the knowledge, that the described forms belong to *one* sphere of development, the same as with the *Aspergillus*; until lately, since the course of development has been minutely studied, the Conidia bearers, Sclerotia and Utricles were considered as a particular and different species of relative sorts of fungi. The first were placed as *Botrytis cinerea, Botrytis vulgaris* etc. in the species *Botrytis*, also called *Polyactis*; the second with other similar bulbous formations, were called *Sclerotia (Scler. durum, Scler. echinatum* are the names of those kinds which mostly resemble those forms which belong here). The former appellative is now applied to distinguish an *organ* or state of development which is present in many species; the Utricles at last belong to that kind which is so rich in variety of form, the cup-fungi, *Peziza, Peziza Fuckeliana,* the sort special to vine-leaves.

There still remains a large quantity of widely spread forms which, though greatly differing in their shapes from the *Aspergillus glaucus* and *Botrytis cinerea,* so far agree with them, in as much as on the Mycelium, which springs from the spores when fully developed, they form successively Conidia (often of two forms) and Utricles. As it is impossible within the prescribed time to allude to these varieties, it will be more expedient to proceed to an example somewhat of another kind than those we have hitherto given.

Mucor stolonifer (Fig. 3), is the name of an important *mildew*, which as a white woolly covering with black stalked heads can be very troublesome on juicy fruits, and does not disdain other organic bodies. Like the fungi already described, its development generally begins with the formation of a Mycelium, which like that of the *Aspergillus* spreads itself through the substrate, but is distinguished from it by its richly branched filaments being without partitions; therefore, they are long ramified Utricles. First in a latter stadium of development, there often appear irregular partitions. From the Mycelium which is spread in the substrate, rise very thick Utricles without partitions, — Stolones, Runners (s) — obliquely in the air, growing to the length of half an inch and more, then bending their points down to the substrate and sending forth from it three kinds of branches. One of these rises perpendicularly with the substrate, is 1—2 lines long, and has at the point a round ampulla, in which spores are formed — Sporangium (p); they can be called *Sporangia bearers*. At the points already indicated, 1—6 appear slightly diverging from one another and rising up from the substrate. The others which spring up near the basis of the Sporangia bearers, become *hairy roots*; they cling to the substrate as richly ramified Utricles, and fasten the Sporangia bearers to it. — The third, generally two in number, take the properties of stolones; at their points the ramifying process is repeated. With sufficient nourishment, it can take place several times. The fungus, by several successive formations of Stolones, spreads itself a few inches over the nourishing substrate, and indeed over the bodies on which the *hair roots* fix themselves. At last, the formation of the Stolones ceases, and the ramification reaches its limits.

The Sporangia bearers swell at their points into round vesicles filled with protoplasma, which are soon separated from their cylindrical bearers by a partition, the form of

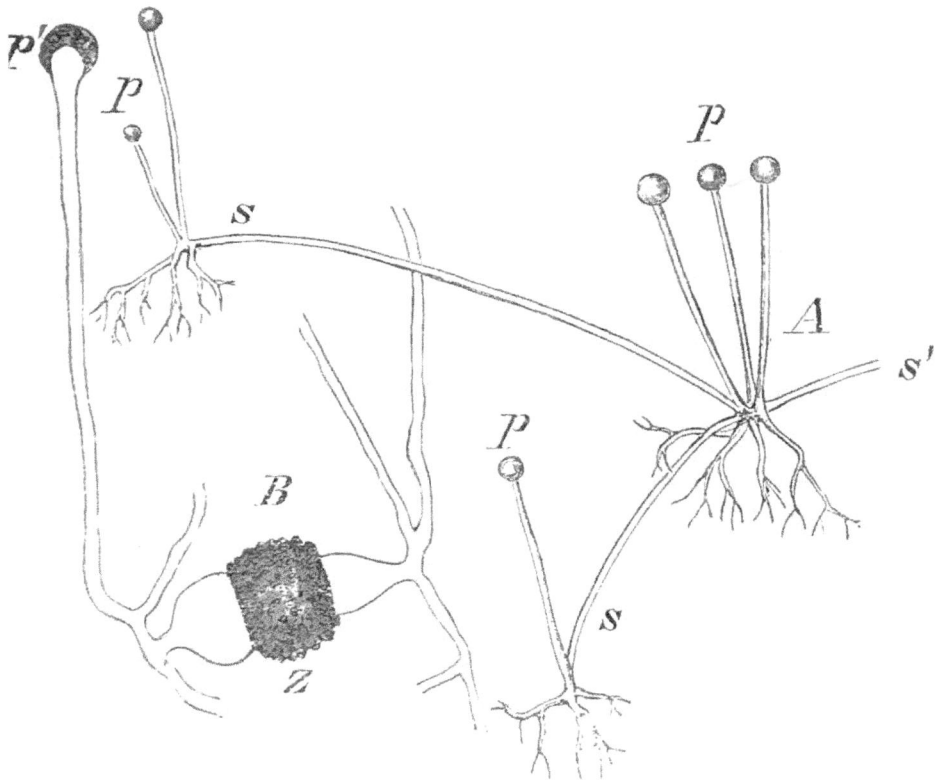

Fig. 3. *Mucor stolonifer.*

A, magnified slightly, *B* mag. about 60 times.

A s' the end of a runner (*stolo*), which has branched out into 3 Sporangia bearers (*p*), a cluster of hair roots and 2 Stolones of the second order (*s*), the ends of the latter with Sporangia bearers (*p*) and hair roots.

B z Zygospores with their bearers. From the filaments which bear them, rises a Sporangia bearer *p'*, the Sporangium of which is shown in its length.

which is not at first flat, but cupola-like and arched towards the upper part, which gives the cavity the shape of a strongly curved Meniscus (p') The place in which it commences (its apophysis) lies rather high over the lower end of the round protuberance. Every meniscus-like cavity is the place in which the spores are formed — *Sporangium, matrix spores.* The whole of the protoplasma which fills it, falls all at once into a great number of polyhedral parts, each of which are soon covered with a particular skin, and become more and more round, in order to produce an equal quantity of spores, the thin partition of the Sporangium, which surrounds them above and below, becomes brittle as the spores ripen, and soon decays by which the spores which are ready to fall away, are set free. The arched partition, which is below the Sporangium, remains standing with the bearers, like a cupola body, on which the traces of the commencement of the outer partition are to be seen in the form of the edge of a ring, and are superfluously called white pillars, *Columella.* After it is ripe, it becomes very strong, as well as the twigs and hairy roots and is of a light brown or violet black colour, the roots have innumerable partitions. — The Sporangia as well as the Columella take this form in a turgid watery situation. When they are withered by evaporation, they fall together in a pileated form or that of a concave-convex lens, in order to resume their original turgescent form, when water is present, a simple but frequently misunderstood fact.

The formation of the Stolons, and the arrangement of the Sporangia bearers are not less definite, they are peculiar to the kind of which we speak, putting aside the form of structure of the single spores. The development of the Sporangium and the spores which it contains, is essentially the same as the numerous kinds of the *Mucor* species. The development of the spores shows, in a superficial observation, a resemblance to that which takes place in the *Asci* of the *Botrytis cinerea* and *Aspergillus* so far as that in both cases the spores in the interior of the matrix are not produced by being unstrung. But a great distinction is, however, present, for in those *Asci* single portions of the Protoplasma which continue in their totality, separate themselves as spores, but by the Sporangium of the Mucor, the whole of the Protoplasma simultaneously divides itself into spores.

In many cases only the already described appearances may be observed on the *Mucor stolonifer*. At the most, it sometimes happens, that direct Sporangia bearers grow out of the Mycelium, which are in every way similar to those produced by the Stolons.

Our fungus has, however, other organs of production: *coupling cells, Zygospores*. Their formation has till now only been observed in the warm summer time, and particularly on sour fruit, but there it exists in masses. To the formation of the Zygospores, the Mycelium sends forth twigs which creep *over* the surface of the substrate, richly branched and often greatly intertwined. At the points, where two twigs cross one another, each one sends forth a short protuberance, the flat ends of which lie close to the ends of the other, which are flattened in the same manner. In this connection, they both grow together to a large spindle-shaped body, each single one representing a club or ninepin, the flat surface of its basis being the part on which both the protuberances come in contact. They attain, therefrom, a thickness which greatly surpasses the filaments which bear them, and a quantity of Protoplasma is collected in them.

Near and parallel with the surfaces which are in contact with one another, a partition springs up in each club, separating the broad end into a particular short cylindrical cell— *coupling cell* — from the other club-like parts, the *bearers* of the coupling cells. The cells of a pair are almost always unequal: the one is as high as it is broad, and the other only half so high. They both melt into *one* through the dissolving of the partition, which served to separate them, the original surface, on which they came in contact. The substance which produces the melting, the *zygospore* (z) now grows still farther, taking a round or ton-like form; it is covered with a thick skin, which at last consists of several layers, and a thick verucose surface, except on the apophysis of the bearers; at the time it is fully formed, it sits, resembling a black cell filled with greasy protoplasma, between the pair of bearers with which it first grew up, and which are at last withering (z). Subsequently, the same filaments which form the *Zygospores*, often shoot forth close to these single spore bearers possessing the properties above described.

With respect to the germination of the Spores and Zygospores in the former, it is very similar to that of the *Aspergillus-Conidia*; on a suitable substrate they shoot forth germinating utricles, these grow to a mycelium from which the afore-described process of formation begins anew.— It is otherwise with the zygospores. If they come into damp ground after they are ripe, they also shoot forth a germinating utricle, which surrounded by the inner layer of the partition breaks the outer one and frees itself — not to develop itself as a mycelium, but to grow, at the expense of the nourishment contained in the Zygospore, rising up and forming itself into one of the above described spore-bearers.

The cycle of development begins anew with the ripeness and germination of the spores, which it has produced. We must here acknowledge, that the germination and germinating products of the *Mucor stolonifer* have not yet been minutely examined, but that the description of the same is given from observations made on other kinds of the species; these are so similar to the *Mucor stolonifer*, and particularly in the formation of the *Zygospores*, that we can say with certainty, that the Zygospores of one species germinate like those of all the others.

If we take a glance back at the fungi, which we have already described, with which a number of others can be classed, they show in common with one another in their course of development, that it reaches its highest point and generally also its end, with the formation of one of its bodies which serves for propagation and which shows, of all parts, of the fungus, the greatest complication in the history of its form and origin.

We call the formation of this body, to distinguish it from other kinds of production, *fructification*. The Zygospores are the organs of *fructification* in the *Mucor*, and the utricles those in the *Botrytis cinerea*. By the fungi which we know more exactly, we find *one* particular fructifying organ by each species. The process of copulation in the *Mucor* is the same as that which is immediately annexed to the reproductive process which is well known among the smaller plants as a particular form. The Utricle of the *Aspergillus* is, as we can only slightly mention here, a reproductive product. The same is undoubtedly true, for those kinds of fungi, which do not belong here, and there are sufficient indications to prove, that the organs of fructification in the *Botrytis* and the great number of fungi, which are furnished with *Utricles*, as well as those fungi which have no Utricles, owe their existence to a sexual process. At any rate so far is certain, that to those *Utricles* and *Asci*, from which the *Spores* are formed, in the manner we have already described, belong only that kind of fructification in the sense of the word as we have just described it.

In the course of their development which reaches its height with the fructification, many fungi — not all —form other unsexual organs, which serve for the purpose of propagation, and which in distinction to fructification, are called *Propagation-organs*, and as in our examples, in consequence of the great number in which they appear, constantly serve to increase the species. The *Conidia* of our examples, and the *Sporangia* and *Spores* of the *Mucor* belong thereto. There are also fungi, which produce several kinds of propagating organs, as the examples will demonstrate, which we will presently give. Each kind shows us, in this respect, its distinctly stamped characteristics. Each of these species is not so much distinguished by *any particular form*, in which it appears, as by its *particular course of development*, in which the productive organs appear in the successive or alternatively changing *different* forms, through a *Pleomorphy* (as the appearance is called), which follows according to definite rules. The most, but we have already said, not all fungi, are with respect to their productive organs *Pleomorphic*. The different organs of a pleomorphic kind do not appear at the same time on the Mycelium, the latter often does not reach the point of its development, *fructification*, when the conditions, under which it vegetates, are unfavourable, and only forms Conidia or other organs of propagation, as the before named examples teach us; it is very seldom, that the opposite case occurs, viz: that the formation of the Conidia is omitted. After these observations, the question, "what is Mildew", can be answered with more decision than it was possible to do at the commencement of this discourse. It is filamentous Mycelia with thread-like fruit-bearers of *Mucor* and fungi, which characterized by utricle fructification, are called *utricular fungi, Ascomyceta*. This definition is appropriate to the greater part of the mildew-fungi, some belong to the pleomorphic species which according to the form of their fructification, are to be classed in other sections, viz: those of *Mucor* and *Ascomycetes*.

From this explanation, it is made clear, that we do not always find the single forms of a species together, and in distinct connection with one another, but that we more frequently find the organs of propagation, than those of fructification. Where such single forms appear, they show (and experience determines that in every case) a similarity with the forms of propagation and fructification, the original connection of which is known, and their morphological signification can with some certainty be determined; an unstringing or loosening apparatus according to that which takes place in the conidia-bearers of the *Botrytis* and *Aspergillus* spores; for example, the conidia-bearers can be regarded as organs of *propagation*, but the spore-utricles (*Asci*) always as those of *fructification*.

From a number of common mildews, only a few forms are known which can be examined according to these principles. A few examples of such *imperfectly known* kinds shall now follow.

If we bring quite fresh horse-dung into a damp confined

atmosphere, for example, under a glass-bell, there appears on its surface (almost without exception), after a few days, an immense white mildew. Upright strong filaments of the breadth of a hair raise themselves over the surface, each of them soon shows at its point a round little head, which gradually becomes black, and a closer examination shows us, that in all principal points it perfectly agrees with the sporangium of the *Mucor stolonifer*. This form of which we speak, is therefore placed among the *Mucor* species, the name of this kind is *Mucor Mucedo* (Fig. 4). Each of these white filaments are its sporangia-bearers. They spring from a mycelium, which is spread in the mist, and is similar to our *Mucor stolonifer*. They appear singly on the Mycelium, and not in clusters on the Stolons. Herein lies the principal difference between this and the other kinds of *Mucor stolonifer*.

Certain peculiarities in the form of the Sporangium, which we will not mention here, and the little long cylindrical spores which, when examined separately, are quite flat and colourless, are characteristics of the species we have just spoken of. If the latter be sown in a suitable medium, for example, in a solution of sugar, they swell out and shoot forth germinating utricles, which quickly grow to mycelia, which bear sporangia (A). This is easily produced on the most various organic bodies, and *M. Mucedo* is therefore found also spontaneously on every substrate, which is capable of nourishing mildew, but on the above named, the most perfect and exuberant specimens are generally to be found.

The Sporangia-bearers are at first always branchless and without partitions. After the Sporangium is ripe, cross partitions in irregular order and number often appear in the inner space and on the upper surface branches of different number and size, each of which forms a sporangium at its point (A). The sporangia which are formed later, are often very similar, but sometimes very different to those which first

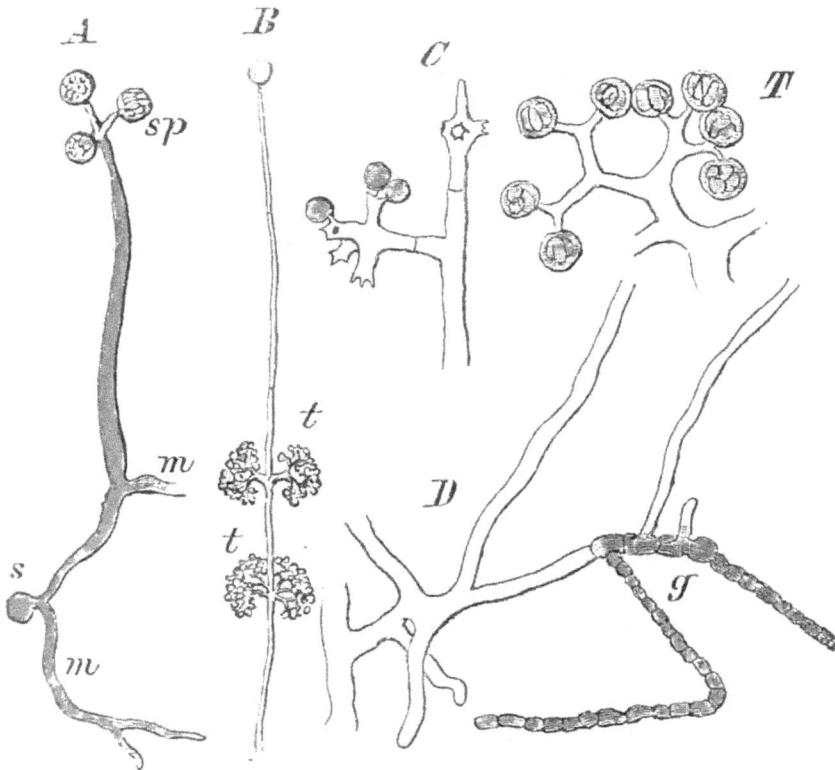

Fig. 4. *Mucor Mucedo.*

A (magnified 100 times) *s* germinating spore from the mycelium filaments (*m*) and from these an inclined sporangia-bearer shoots forth
with three sporangia (*sp*).

B (slightly magnified) The point of a sporangium-bearer with a large sporangium on the top and two whorls of sporangiolum bearing little
twigs (*t*).

T The end of such a twig with ripe sporangiolum (magn. 200).

C The point of the branch of a conidia-bearer, 3 conidia are on it, the
others have fallen off (magn. 300).

D (magn. 190) A piece of a mycelium filament, from the branches of which
two (*y*) germs are devided into innumerable short members, grown in
a solution of sugar.

appeared, because their partition is very thick, and does not fall to pieces, when it is ripe, but irregularly breaks off, or remains entire enclosing the spores, and at last falls to the ground, when the fungus withers. The cross partition which separates the Sporangia from its bearers, is by those, which are first formed (which are always relatively thicker Sporangia) very strongly convex, and do honour to the name of *colum*, those which follow later, are often smaller, and in little weak specimens much less arched and sometimes quite straight.

After a few days, similar filaments generally show themselves on the dung between the already described Sporangia-bearers, which appear to the naked eye to be provided with delicate white frills (B). Where such an one is to be found (t), two to four rectangular expanding little branches spring up to the same height round the filament. Each of these, after a short and simple process, branch out into a furcated form; the same furcation is separated in several orders in such a manner, that the ends of the branch at last so stand together, that their surface forms a ball. Finally, each of the ends of a branch swells to a little round sporangium, which is limited by a partition (called sporangiolum to distinguish it from the large ones), in which some, generally four spores, are formed in the manner already known (T).

When the Sporangiola are alone, they bear such a peculiar appearance with their richly branched bearers, that they can be taken for something quite different to the organs of the *Mucor Mucedo*, and were formerly not considered to belong to it. That they really belong to the *Mucor Mucedo*, is shown by the principal filament which it bears, not always, but very often ending with a large Sporangium, which is characteristic of the *Mucor Mucedo* (B); it is still more evident, if we sow the spores of the Sporangiolum, for as it germinates, a mycelium is developed, which near a simple bearer can form large sporangia and those from Sporangiola; the first always considerably preponderating in number, and very often exclusively. If we examine a large number of specimens, we find every possible middling form between the simple or less branched sporangia-bearers and the typical sporangiola-frills, and we arrive at last at the conclusion, simply to place the latter among the *varieties*, the form which the sporangia-bearer of the *Mucor Mucedo* shows like every other typical organic form within certain limits.

It is certain that we may expect according to the resemblance in the formation of Mycelia and Sporangia bearers with the *M. stolonifer*, that *M. Mucedo* also produces *Zygospores* as a form of *fructification.* If I do not err, these have been seen, and are very similar to the *M. stolonifer*; but they are not known with sufficient certainty.

On the other hand, a propagation organ, differing from those of the Sporangia and their products, belongs to the *M. Mucedo*, which fails in the *M. stolonifer*, and in the above used terminology can be called Conidia respectively Conidia-bearers. On the manure (they are very seldom on any other substrate), the latter appear at the same time · or generally somewhat later than the sporangiola-bearers, and are not unlike those to the naked eye. In a more accurate examination they appear different; a thicker, partitionless filament rises up and divides itself generally three-forked at the heigth of $1'''$, into several orders. The forked branches of the last order bear under their points which are mostly capillary, short erect little twigs and these with the ends of the principal branches

articulate on their somewhat broad tops several spores and
Conidia near one another quite in the same manner as is described by the *Botrytis cinerea*; about 15—20 are formed at
the end of each little twig (G). The peculiarities and varia-
tions, which so often appear in the ramification, can here remain undiscussed. After the articulation of the Conidia, their
bearers sink together by degrees, and are quite destroyed.
The ripe Conidia are round like a ball, their surface is
scarcely coloured and almost totally smooth (C).

After that which has been said, the idea, that our Conidia-bearers belong to the sphere of development of *M. Mucedo*
lies very far and Berkeley was certainly right, when he (after
a simple examination of the complete single form) considered
them related to the *Botrytis cinerea*, and called them after
the discoverer *Botr. Jonesii*. Why then do they belong to
the *Mucor*? That they gregariously come forth, is as little
proof here of an original relation to one another, as elsewhere. Attempts to prove, that the Conidia and Sporangia-
bearers originate on one and the same Mycelium filament, as
by the *Aspergillus* may possibly once succeed. Till now, this
has not been the case, and he who has ever tried to disen-
tangle the mass of filaments, which exuberantly covers the
substrate of a *Mucor* vegetation, which has reached as far as
to form Conidia, will not be surprised, that all attempts have
hitherto proved abortive.

The suspicion of the connexion, founded on the gregarious springing up and external resemblance is fully justified,
if we sow the Conidia in a suitable medium, for example, in
a solution of sugar. They here germinate and produce a mycelium, which exactly resembles that of the *Mucor Mucedo*,
and above all they produce in profusion the typical Sporangia

of the same on its bearers. The latter are till now alone, reproductions of Conidia-bearers, and have never been observed on Mycelia, which have grown out of Conidia.

These phenomena of development appear in the *Mucor Mucedo,* when it dwells in a damp substrate, which must naturally contain the necessary nourishment for it, and is exposed to the athmospheric air. Its Mycelium represents at first, as we have already observed, strong branched utricles without partitions, the branches are of the higher order mostly divided into rich and very fine pointed twigs. In old Mycelium and also in the Sporangia-bearers, the contents of which are mostly used for the formation of spores, and the substrate of which is exhausted for our fungus, short stationary pieces filled with Protoplasma, are very often formed into cells through partitions in order to germinate spores, that is, grow to a new fruitful Mycelium. These cells are called *gems*, *brooding cells*, and resemble such vegetable buds and sprouts of foliaceous plants, which remain capable of development after the organs of vegetation are dead, in order to grow under suitable circumstances to new vegetating plants; as for example, the bulbs of onions etc.

If we bring a vegetating Mycelium of the *Mucor Mucedo* into a medium which contains the necessary nurishment for it, but is enclosed from the free air, the formation of Sporangia takes place very sparingly or not at all, but the formation of *gems* is very abundant. Single intersticial pieces of the twigs, or even whole systems of branches are overfilled with a rich greasy Protoplasma, the short pieces and ends are bounded by partitions which form particular, often ton-like or globular cells, the longer ones are changed through the formation of cross partitions into chains of similar cells (Fig.

4, D by g), the latter often attain by degrees strong thick walls, and their greasy contents often forms itself into innumerable drops of a very regular globular form and of equal size. Similar appearances show themselves after the sowing of spores, which are capable of germinating in the medium already described, from which the air is excluded. Either short germinating utricles shoot forth, which soon form themselves into rows of *gems*; or the spores swell to large round bladders filled with Protoplasma, and shoot forth on various parts of their surface innumerable protuberances, which fixing themselves with a narrow basis, soon become round vesiculate cells, and on which the same sprouts which caused their production, are repeated — formations which remind us of the fungus of fermentation, which we will describe further on, and which will be called *globular yeast* (*Kugelhefe*) (See Fig. 7, A, page 60). Among all the known forms of *gems*, we find a variety, which are intermediatory; all of which show, brought into a normal condition of development, the same proportion and the same germination as those we first described.

This ends the series of forms which have been minutely examined in the *M. Mucedo*. I will now only remark, that the name *Mucor racemosus* designates a form, which is generally distinguished from the strong *M. Mucedo* by the smaller size of every part, and that on their Sporangia-bearers appear numerous scattered short side branches, which also bear a Sporangium. The *M. racemosus* dwells in the most different substrates which are capable of mildewing, as of fruits, stale food etc. If it be really distinguishable from the *M. Mucedo*, we will not here determine. If is particularly disposed to the formation of *gems*, and in consequence has played many a trick on the fungi enthusiasts. They are such as are celebrated as particular fungi, *Urocystis Cholerae,* because they appear in cholera dejections, partly alone, partly mixed with other things, and which, as they are presumptively new, cannot be any-thing else, than the cause of the Asiatic cholera, the "*Cho-lera fungi*".

To the imperfectly known forms of Mildew belongs un-fortunately also that, which is most common, the mildew par excellence, *Penicillium glaucum Lk.* (= *P. crustaceum Fries*) Fig. 5, a kind which seldom ever fails in any place where mildew can be produced, and often totally takes possession of its terri-tory in the form of a thick short coating, first white, then soon covered with a greyish-blue and dirty green-ish-grey dust. The Mycelium of that fungus produces generally tolerably stiff cross partitions, richly branched cylindrical filaments, but which also appear wavy and sometimes show cells swollen to broad bladders, and standing in rows behind one another. From some of the cells of the My-celium, also from the last mentioned bladders, shoot up in the air, as per-pendicular branches Conidia-bearers, equally strong as the filaments of the Mycelium. The articulation of these follows in the most simple form. First branchless and divided by cross partitions into procumbent cylindrical cells, they soon cease to lengthen and their upper cell becomes subulated. From the upper part of the nearest under cell,

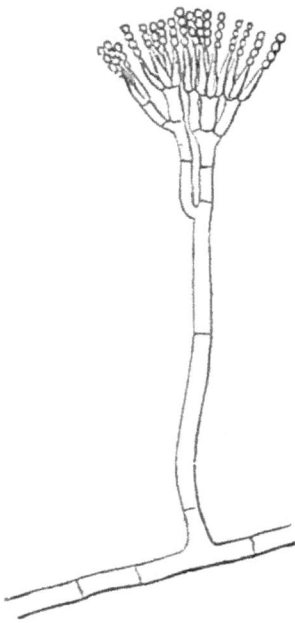

Fig. 5. *Penicillium glaucum.* A Conidia-bearers shooting forth from a Mycelium fila-ment. Magnified 375 times.

1 or 2—3 little twigs spring up at the same time, opposite to one another, which raise themselves close to and nearly parallel with the cell at the point, and like it consist of *one subulated cell.* The points of the three *subulated cells* are of the same height. In strong specimens, branches shoot from the upper and of the third and even fourth uppermost cell, which raise and ramify themselves in the manner which we have already described (Fig. 5). According to the strength of the specimens, the branches of the different rows stand alone or two together, opposite one another. Those in the last row are always of a subulated form and raise themselves almost parallel to the same height with one another. At the point of a fruit-bearer, there are one or more rich clusters of subulated twigs, according to the size of the specimen. Each of these articulate successively, exactly like the *Sterigma* of the *Aspergillus*, a long row of spores — Conidia — the articulation keeps the same place in every member of the cluster of twigs, and these at last bear on their tops a quantity of rows of spores of a parallel and equal height; the single spores of which after they are ripe, are easily reduced to dust. The spores are small, globular and smooth, and when seen in quantities, are of a greyish-blue colour, which they even confer on older *Penicillium* coverings; but when they are examined singly, the colouring is so weak, that it is often scarcely to be recognized.

A pretty little variety of the *Penicillium*, which was formerly known under the name of *Coremium glaucum*, is only distinguished from it, by the fruit bearing filaments, which raise themselves from the substrate, and closely intertwined. form thick clusters in the shape of a sheaf about 1 line high, and on the top of which the development of the Conidia follows.

Another form, which according to the above description belongs here as a variety which was first found in the acoustic ducts of a patient, and also appears elsewhere, resembles the Conidia-bearing *Aspergillus* so far, as its utricles are thick, simple and have at the end bladder-like globular swollen filaments and the swelling at the point is thickly covered with radiately diverging Sterigmata. These have an oval form, shoot forth from their points several little diverging twigs consisting of one cell and standing close to one another, and each one articulates; as in the form of *Penicillium* a chain of globular smooth spores.

Those spores produced on the bearers resembling those of the *Aspergillus*, as well as those of the common form easily germinate in water, after they are ripe, or in watery solutions and damp substrate, in which they shoot forth germinating utricles, which soon grow to a Mycelium, which produces Conidia. And indeed, till now *all* the spores, which have been sown for the purpose of experiments, have only re-produced the common clustering Conidia-bearers, *not* those resembling the Conidia-bearers of the *Aspergillus*.

From the last mentioned fact, it appears from Cramer's statement, that the described forms belong to *one* species. Further articles of development have not, till now, been discovered. *Penicillium glaucum* has an unmistakeable resemblance to the *Aspergillus glaucus* in the manner of the articulation of its spores, and particularly in those fruit-utricles (fruit-bearers), which are designated as similar to those of the *Aspergillus*. We can therefore without hesitation regard it as the Conidia-form of a species of fungi. From this cause, and because they are so often found together, the idea has arisen, that *Penicillium* must be one of the forms belonging to the *Aspergillus glaucus*, and this seems very plausible, as by a great number of fungi there appear, as we have already stated two sorts of Conidia, and sometimes many intermediate forms. If this surmise were correct, then our *Penicillium* would belong to the form of a well-known *Ascomycetes*. Until now, no one has succeeded in authenticating this, and it must still remain doubtful, when the question, to which form this, the commonest of all mildews, belongs, will be settled,— Attempts have been made to bring the *Penicillium glaucum* in the series of form of the *Mucor Mucedo*, the *Oidium lactis* and others, the result of which we will speak of later.

In concluding our observations on certain examples, a few words must be said on a form, which must, till now, remain unmentioned, when we speak of the common form of mildew. It has the name of *Oidium lactis*, because it very frequently (but not constantly) appears on sour milk, and is found besides on the most various kinds of substrate; very often, for example, on the excrements of animals, and even on those of men; it was discovered in those of cholera patients, considered to be something new, and called *Cylindrotaenium Cholerae*. From no greater fact than this, the discoverer of the *Cylindrotaenium* concluded, that it was the *"cholera fungus"*, the bearer of the *cholera contagion*.

The common form of this fungus (Fig. 6), always appears to the naked eye white as snow, and where it is richly covered, in the form of a thick downy crust. With the microscope, we can recognize a cross partitioned richly branched Mycelium (m) resembling that of the *Penicillium glaucum*, often distinguished from it by the great tenseness of its ramification. From the Mycelium, which is spread through the substrate, branches spring forth and raise themselves in the air as Conidia-trainers (Fig. 6). They attain to different lengths, mostly less than half a Mm., and then divide themselves,

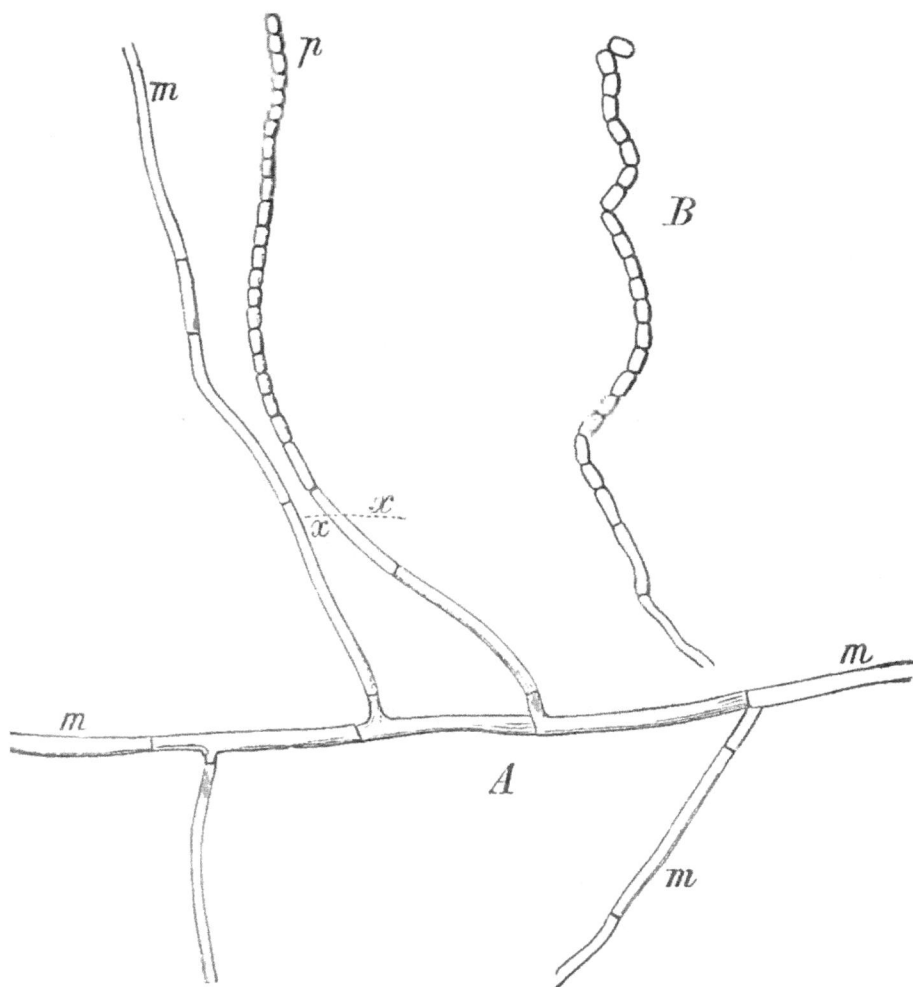

Fig. 6. *Oidium lactis* (magnified about 350 times).

A branched Mycelium filaments horizontally stretched out in the fluid substrate *m—m.* with *a* (by the line *x—x*) branch rising crookedly in the air and divided by cross partitions into a chain of cylindrical spores.— *B* a chain of spores beginning to separate its members from one another.

with the exception of a short under piece, in their whole dimension, into a row of cylindrical members, which are once or twice as long as they are broad (p). Each of these represents a Conidium. Soon after they are formed, they separate from one another; at first imperfectly and so, that the chain appears broken in a zigzag manner (B), but soon after, it falls entirely to pieces. When cultivated in a fluid substrate, and sometimes in poor specimens, such Conidia cells are formed from twigs within the fluid. This undoubtedly takes place, but not so often as it appears in a superficial observation, in which we find in the fluid great quantities of Conidia, which have *fallen off* after they are ripe, and here and there loosely united themselves in rows. The Conidia easily germinate in suitable substrate, and from them originates a Mycelium, which forms Conidia of the same shape. Those who have read the above description or something more of the literature of fungi, will consider the described *Oid. lactis* as a *single form* of a greater *series of forms*. To these belongs, as far as our examinations reach, another form of Conidia-bearers, which has not till now, been described, and the existence of which can here only be mentioned. Other members of the same are not known. It is true, that many statements assert the contrary; for, many place the *Oidium lactis* simply as a form of the *Mucor Mucedo* or the *Penicillium glaucum*, or as belonging to both; unfortunately without being able to give any certain proof of what they have affirmed.

It will not appear strange to any one, that among the common forms of Mildew, there are some, of which it has been impossible to gain the knowledge, to which form they belong, as to trace out such a complicated course of development, as that of the fungi of which we speak, requires so much care and time, and the investigation of these things is of recent date; but on the contrary, the reader will with reason wonder, that it has often been said, that the forms like *Penicillium*, *Oidium lactis*, — and many others could be named — belong to the sphere of development of such and such a form, but the assertion is groundless. An impartial observer will think, we can see with or without a microscope, whether an organism develops itself in this or that manner, whether the same be called tree, moss or fungus; we must be prepared to meet unsuccessful experiments, open questions in this as in every other case, also differences of opinion in points of detail, but not diametrically opposite views on the connexion of sharply characterized forms like those we have named. That such a difference exists, is explained in a simple, perhaps for the impartially instructed, in an unexpected manner.

In order to determine, whether an organic form, an organ or an organism, belongs to the same series of development as another, or, that which is the same, is developed from it or vice versa, there is only one way, viz: to observe how the second grows out of the first. We see the commencement of the second begin as a part of the first, perfect itself in connection with it, and at last it often becomes independent, but be it through spontaneous dismembering from the first, or that the latter be destroyed and the second remains. Both their disunited bodies are always connected together in *organic continuity*, as parts of a *whole* (single one); that can cease earlier or later. By observing the organic continuity, we know, that the apple is the product of development of an apple-tree, and not hung on it by chance; that the pip of an apple is a product of the development of the apple, and that from the pip an apple-tree can at last be developed; that therewith all these bodies are members of a sphere of development or form. It is the same with every similar experience of our daily life. That where an apple-tree stands, many apples lie on the ground, or that in the place where apple-pips are sown, seedlings, little apple-trees, grow out of the ground, is not important to our view of the course of development. Every one recognizes that in his daily life, because he laughs at a person who thinks a plum which lies under an apple-tree, has grown on it, or that the weeds, which appear among the apple-seedlings, come from apple-pips. If the apple-tree with its fruit and seed were microscopically small, it would not make the difference of a hair's breth in the form of the question or the method of answering it, as the size of the object can be of no importance to the latter, and the questions which apply to microscopic fungi, are to be treated in the same manner. If it then i. asserted, that two or several forms belong to a series of development of one kind, it can only be based on the fact of their organic continuity; on this is founded that which we have before said respecting the sphere of form of the *Aspergillus, Botrytis, Mucor* etc. This proof is more difficult than by large plants, partly because of the delicacy, minuteness and fragility of the single parts, particularly the greater part of the Mycelia; partly because of the resemblance of the latter in different species, and therefore follows the danger of confusing them with different kinds, and finally, partly in consequence of the presence of different kinds in the same substrate, and therefore the mixture not only of different sorts of Mycelia, but also, that different kinds of spores are sown. With some care and practice, these difficulties are in no way insurmountable, and they must at any rate be overcome, the organic continuity or non-continuity must be cleared up, unless the question respecting the course of development and the series of form of special kinds be not entirely laid on one side as insolvable.

Simple and intelligible as these principles are, they have not always been acted upon, but partly neglected, partly expressly rejected; not because they were considered false, but because the difficulties of their application were looked upon as insurmountable. Therefore, another method of examination was adopted, the spores of a certain form were sown, and sooner or later they were looked after to see what the seed had produced,—not every single spore, but the seed en masse; that is, in other words, what had grown on that place where the seed had been sown. As far as it relates to those forms which are so widely spread, and above all grow in conjunction with one another, — and that is always the case in the specimens of which we speak — we can never be sure, that the spores of the form which we mean to test, are not mingled with those of another species. He who has made an attentive and minute examination of this kind, knows, that we may be sure to find such a mixture, and that such an one was there, can be afterwards decidedly proved. From the seed, which is sown, those spores, for which the substrate was most suitable, will more easily germinate, and their development will follow the more quickly. The favoured germs will suppress the less favoured, and grow up at their expense. The same relation exists between them as between the seeds, germs and seedlings of a sown summer plant, and the seeds, which have been undesignedly sown with it, only in a still more striking manner, in consequence of the relatively quick development of the mildew fungus. Therefore, that from the latter a decided form or a mixture of several forms is to be found sown on one spot, is no proof of their generic connexion with one which has been sown for the purpose of experiments, and the matter will only be more confused, if we call imagination to our aid and place the forms, which are found near one another, according to a real or fancied resemblance, in a certain series of development. — All those statements on the sphere of form and connexion, which have for their basis such a superficial work, and are not based on the clear exposition of the continuity of development as by the origin of the connexion of the *Mucor* and *Penicillium,* *Oidium lactis* and *Mucor,* *Oidium* and *Penicillium,* are rejected as unfounded.

A source of error, which can also interfere in the last named superficial method of cultivation for experiments is, viz: that heterogeneous unwished for spores intrude themselves *from without* among the seed which is sown; but that has been until now quite disregarded. It is of great importance in practice, but in truth, for our present explanation synonymous with that we have already said. Those learned in the science of this kind of culture lay great stress on its importance, and many apparatuses have been constructed, called *"purely cultivating machines"*, for the purpose of destroying the spores which are contained in the substrate, and preventing the intrusion of those from without. The mixture in the seed, which is sown, has of course, not been obviated. These machines may, perhaps, in every other respect fulfil their aim, but they cannot change the form of the question, and the most ingeniously constructed apparatus cannot replace the attention and intellect of the observer.

As we have become acquainted with a number of mildew forms, we can now proceed to take into consideration their general characteristics, particularly their physiological. Any example, chosen at pleasure from the whole number, will serve as a starting point. Those we have already brought forth, are particularly adapted to this purpose, because physiological examinations have been more specially made of those than of the other fungi.

Whoever has examined the vegetable kingdom and the results of botany, will recognize that the mildew fungus shows like every other kind, its peculiarities in form and development, but that in the principal points of its organization, it perfectly agrees with the typical members of the whole of the vegetable kingdom, and is a plant like all the others. Even the appearance of the *fruit pleomorphy* does not belong specially to the fungi; similar and sometimes very strictly comparable phenomena are commonly found in the great series of cryptogams, those are plants which are propagated by other means than by blossoms, or in other words, plants, the stamens and pistils of which are not visible.

It is perfectly agreed, that the main point between the fungi and other plants is in complete accordance, viz: the combination of organic combustible substances and incombustible mineral or ash ingredients; as by the latter, it is also by the former, that the mixture of substances is different, each according to its species and organ.

Thus we see, that the mildew fungus is essentially dependant on the same conditions of vegetation as other plants. Each of the normal functions of every species requires a cer-

tain degree of warmth, many, if not all, a certain amount of
light, and hydrogen and oxygen is absolutely necessary to its
normal vegetation. From the homogeneousness of the material
combination it results, that in the whole the same elements of
combustible organic and incombustible substances must be supplied to the nutritive matter as in all other plants. The fungus shows one peculiarity — but also all other plants besides
fungi, which are not chlorophyll — with respect to the absorption of the combustible substance, which forms its nutritive matter. The chlorophyll plants generally absorb this
(as we stated in the first part of our lecture), in a highly
oxydated *inorganic* combination; indeed (if we omit sulphur),
their hygrogen, oxygen and nitrogen in the form of water,
ammoniac and nitric acid in combination, and their carbon
in the form of carbonic acid. Their assimilation, particularly
the transposition of carbonic acid into the complicated carbonic combination, which we call organic, is requisite to the
green colour of the Chlorophyll. As the fungi are always
without this latter substance, they require organic combinations, which have been already prepared to nourish them and,
according to the examinations partly carried out and partly
incited by *Pasteur*, it is especially carbon, which must be
supplied to them. In a medium, which contains every other
nutritive matter in a suitable form, but carbon only as carbonic acid, we do not find any increase in the growth of a
fungus; that only follows, when it is supplied with an organic carbon combination. For the fungus used for the purpose
of experiments, very heterogeneous combinations replace one
another, as sugar, glycerine, organic acids, tannin etc. Nitrogen can, on the contrary, as far as the present examinations
reach, be taken up in the form of inorganic (ammoniac, ni-

tric acid), as well as organic combinations; if also, (as it is asserted), as free nitrogen gas from the air, must be more closely examined and can here remain undecided. That all nutritive matter must be dissolved, or at least, when it comes into contact with the organ, (Mycelium), which absorbs the nourishment, in a dissoluble form, is appliable to the fungus, for the same reasons as to all other plants.

With the process of vegetation of each fungus, there is also connected that of constant respiration, breathing, that is, the absorption of oxygen, at the same time that it exhales carbonic acid; a condition which entirely agrees with that of all plants and parts of plants, which are not green. Other gas secretions may here remain unnoticed.

The oxygen which is used for respiration, with the exception of some cases, which we will name later, is absorbed from the air; the nutritive matter, on the contrary (without regard to the doubtful assertions given about the absorption of nitrogen gas), is taken up by the Mycelium from the ground. Thus its material combination is of the greatest importance for the nourishment of the fungus. There are sufficient indications, that its physical nature, aggregate condition, consistence, if we may be allowed the expression, and moisture is most important to many kinds, even if the examinations have not been carried very deeply.

If it still remains to follow more especially and to discover more distinctly the differences of the vegetative condition of single species, that which is known and which we have already communicated, is sufficient to explain, why certain kinds of Mildew are exclusively or particularly found in certain substrates — for example, *Botrytis cinerea* almost spontaneously and exclusively on withering but not quite disorganized parts of plants; — why others, as *Aspergillus*, *Penicillium*, *Mucor stolonifer*, flourish on very different kinds of substrate, and why those sorts, which are less particular than the last mentioned, so often appear together, whether it be that they share their territory more or less equally, or that the one supplants or suppresses the other. According to a similar experience, which every one makes daily on large plants, cultivated plants and weeds, the explanation is so clear, that it is not necessary to enter into further details.

An hypothesis is necessary, if we wish to explain the spreading, gregarious mutual supplanting of the mildew form; like the similar phenomena in the other parts of the vegetable kingdom, from the quality of the substrate and other conditions of vegetation; viz: this, that the germs, spores, Conidia etc. of single species, which are capable of development, reach the substrate like the seeds of plants and weeds in the fields and meadows.

Every authenticated fact shows the verity of thisassertion. If we look back at those forms which we have singly described, and observe a patch of mildew, which is rather thick, we find, the fruit-bearers (Mycelia) are close to one another, which is sufficient to demonstrate the immense fertility of the mildew fungus and its enormous productivity in Conidia and other propagating cells, and he who likes large numbers, can easily calculate the amount of Conidia produced on a little surface. A glance at the description we have already given, also shows further, that the large number of spores which are produced, are, immediately that they are ripe, generally set free and fall to dust, and in consequence of their lightness can and must be easily carried away and disseminated by every current of air, and by every thing which moves from one place to another. In accordance with this, the ingenious experiments of *Pasteur*, in which a quantity of air is made to pass through balls of pure wool or similar substances, show, that spores of the common mildew form are deposited in the wool, and we often find them in prodigious quantities affixed to the surface of various solid bodies, whether they be, according to the common designation, "dusty" or not. That through this universal diffusion of the spores the common form of mildew must reach substances favourable to their germination and further development, when these are accessible, is plainly to be understood; and it can be asserted that there is no case in which mildew appears, that through an early investigation and careful attention, we find its origin is assignable to its spores (Conidia etc.). The objection is often raised, that mildew is often found in the interior of uninjured eggs, nuts etc. where it was impossible for the spores to reach. However, the spores settle on the surface of such substances, a slight moisture causes them to germinate, and the germinating utricles, respectively those from a full grown Mycelium filament, penetrate into the enclosed space, piercing through (as is proved by minute observation) the closed vessel, egg-shells, and even the hard and solid membrane of fruit-stones, nuts and woody fibres. Every apparent objection is thus removed.

Phenomena such as we have just mentioned, together with the general appearance of mildew in substances which have begun to decompose, were often until recently brought forward to support the view, that the mildew fungus proceeds from the so-called *primeval production, parentless production*, that is, they do not spring from the spore, which originated in their parents, but from germs, which from the substances taken possession of by mildew, crystallize as organized decomposing products. We do not know, until now, any well ascertained fact, which confirms this view of the mildew fungus. Rather these facts and many careful experiments tend to show, that mildew and fungi, with respect to their origin, differ in no way from other plants. The universal question on the parentless origin of organism in the different periods of production will not be discussed here; we may casually remark, that fungi are not the proper objects of examination, wherewith to decide this point.

In opposition to the view, which considers the mildew fungi a product of decomposition, they show themselves much more as we have already asserted, as *producers,* powerful *exciters* of the process of decomposition, which appears on dead organic substances. Positive decompositions appear, when a certain fungus settles itself, be it spontaneous or after it has been sown, on a substance capable of decomposing, and which must of course contain the necessary conditions of vegetation, of which we have before spoken; they cease, when the fungus is kept away from it, and is put a stop to when the fungus dies; it is, therefore, an effect of its life and vegetative process.

We divide these decompositions into two classes: *putrefaction* and *fermentation.*

The former appears, when a mildew fungus vegetates on the free surface of its substrate, and has an unlimited supply of oxygenated air. By taking up oxygen from the air, the organic substance of the substrate is changed into carbonic acid, water and ammoniac. Carbonic acid and water result from the combination of the carbon and oxygen of the substance in the substrate, (which does not contain any nitrogen), with the oxygen, which has already been absorbed, or, as it is usually expressed, from an oxydation or combustion, of the same. The latter does not take place, or at least very slowly, when *caeteris paribus* oxygen is present in great profusion, but the fungi vegetation must be kept away. Hence it shows, that the latter absorbs the oxygen from the air which surrounds it, and transfers it to the combination in the substrate in a manner, which must yet be more clearly ascertained. A relatively smaller quantity of the latter will of course be used as a material to increase the substance.

The same fungus, which by an unlimited supply of oxygenated air causes putrefaction, can, by limiting or excluding the atmospheric oxygen from it, produce fermentation in the substrate; that is, a disuniting of the organic combinations, which are present, into others which are more simple, but different to the products of putrefaction.

As an example, we will give the result of a work by *van Tieghem* on the vegetation of *Tannin*, the active principle in the *gall apple*. It has long been known, that on exposure to the air and covered with mildew, in absorbing water it decomposes into *gallic acid* and *sugar*. The mildew fungus which *van Tieghem* here observed, are *Penicillium glaucum* and *Aspergillus niger*, a species related to the *A. glaucus*, but entirely distinct from it.

If we take a solution of Tannin, in which the nitrogeneous and mineral substances, which are necessary to the vegetation of the fungus, are mixed, and exclude from it all atmospheric oxygen, it remains unchanged, whether the spores of fungi be sown in it or not. The same effect is produced by the admittance of a quantity of oxygen and keeping the fungus away. If we sow the fungus in the solution, when oxygen is present, the spores germinate and the fungus begins to vegetate. If we let it grow on the free surface of the solution with an unlimited supply of air, it developes itself immensely into carbonic acid and water, by burning the tannin. If we limit the supply of air after the vegetation has commenced, and take care, that the Mycelium is immersed in the solution, it grows more slowly and all the Tannin is changed into gallic acid and sugar, a relatively small quantity of the latter is also here used to increase the growth of the substance of the fungus, which is free from nitrogen.

To explain this altered effect of the enclosed fungus on the substrate, to which an unlimited free current of air has access, it is necessary, above all, to remember, that it can be experimentally affirmed, that not one of the constituent parts of the fungus are given up to the solution, in order to form a chemical combination and thus to produce the decomposition of the Tannin. The decomposition of the freed substances follows much more on the slowly progressing, but still progressing fungi vegetation.

As we now see, that under the same circumstances, the same living fungus, in the same substrate, at one time vegetates luxuriantly and causes putrefaction, and at another produces by slower vegetation fermentation, solely according to the supply of atmospheric oxygen, whether it be limited or unlimited, we must seek for the cause of the different effects in the latter point. *Pasteur* has formed a very clever hypothesis to show in which manner we can form a more distinct idea of this, although it does not refer to the case in question, it is quite analogous to it. He says the fungus must absorb oxygen, when it vegetates. If this be suffiiently supplied to it by the air, then it absorbs it, vegetating luxuriantly and burning the substrate; if it do not find it free and in the air, then it draws it from the combination of which the substrate is composed, and gives the impulse to the further decomposition of this combination.

It is, therefore, intelligible, that a fungus which grows on an organic substance, can produce, at the same time, putrefaction and fermentation; the former, the surface of which is exposed to the air; the latter, where its Mycelium has penetrated into the depth of the susbtrate, which is shut out from the oxygen in the air. In every putrefaction, with the products of putrefaction, appear those of fermentation, which at last by continued fungus vegetation, putrefy also, as is proved by the gallic acid and sugar of our example.

Undoubtedly many or all of the mildew fungi which are known as *exciters* of putrefaction, can produce fermentation under the conditions already described, of course different kinds of decomposition in different substrates. In thegreater part, these effects have not yet been studied. An example can, however, be given in the above described *Mucor Mucedo* and *racemosus*. On a suitable substrate growing in the air, it produces putrefaction by a luxuriant formation of spores. Sown in a fermenting solution of sugar, with a limited supply of air, it produces a Mycelium immersed in the solution, and luxuriant gems, but no spore-leaves, and causes alcoholic fermentation, decomposition of the sugar dissolved in alcohol and carbonic acid.

The alcoholic fermentation, which we employ so much in practical life, is not generally produced by *Mucor*. We owe it, chiefly, to another organism, the *yeast*, the observation of which leads us to the second part of our subject, fermentation and first to the yeast fungi.

Yeast, with its botanical name *Saccharomyces cerevisiae,* also called *Cryptococcus, Hormiscium cerevisiae,* appears in fermenting fluid as a fine uniform clay-coloured mass. This consists of an immense number of cells of plants (Fig. 7, B) which, when in their full growth, are round or oval, and about $\frac{1}{100}$ Mm. large, in every other way, like the cells of the fungus which we have already described, colorless, fragile and containing in the interior of its protoplasma one or more vacuoles filled with water. The single cells are mingled together quite freely in the yeast, or very slightly bound to one another. If they be brought into a fluid, which is favorable to their vegetation, for example, a solution of sugar, capable of producing an alcoholic fermentation, its increase begins by budding, that is, every cell shoots forth a little protuberance like that of the fruit-bearers of the mildew fungus, which grows in the form and to the size of its mother-cell, and then divides itself into an independent cell. On the cell in the second order, the same process is repeated, and

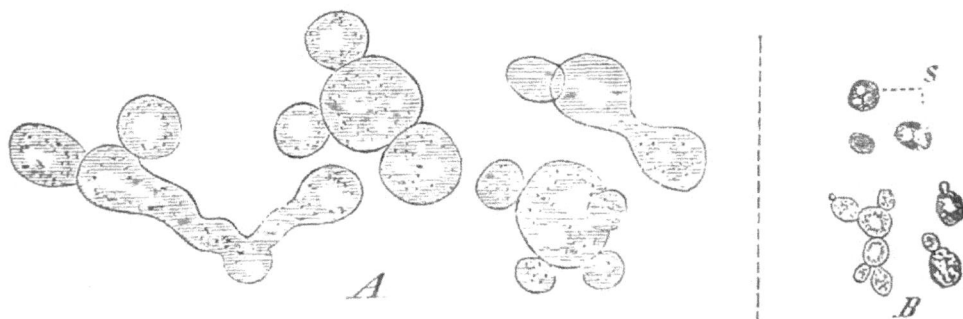

Fig. 7. Magnified 375 times.

A Sprouting gems of a round and irregular long form of *Mucor Mucedo* produced in a solution of grape-sugar.

B Yeast, *Saccharomyces cerevisiae* s. *asci* almost ripe, the one with 3, the other with 2 spores.

also further in an unlimited number of orders or generations. The budding takes place on any part of the cell, often
at the same time or successively at several points. After its
limits are defined, the new shoots separate themselves from
their genitor; we then only find 2 or 3 cells connected together. Or, sometimes by carefully maintained fermentations,
several generations remain united to one another in branched
rows of round members, which can be compared to the
branched short filaments of the fungus. By means of these
sprouts, the yeast increases so strongly, when it is brought
into the fermentable fluid, and it is the budding cells sus-
pended in the fluid, which cause it to be turbid and which
are again deposited, after the fermentation is at an end.

The increase by germination partially resembles in form
the growth of the branched filament of the fungus, and partly
the formation of Conidia by articulation, and can, therefore,
be considered as an intermediate form. It is the only form
of development of the *Saccharomyces*, which has been observed in fermentable solution and during the course of fermentation. If we bring living cells of yeast out of the fluid
or the moist surface of a succulent part of a plant, for example, a piece of a carrot, the sprouting goes on slowly for
some time, and entirely ceases after some days. About the
6th day, we remark, how some of the cells wither and
others become larger; the greater part of the latter form
spores in their inner space through the free formation of
cells, like those of the *Ascus*, and then becoming thicker at
the cost of the Protoplasma, at last entirely fill the membrane of the *Ascus*. We can produce the same phenomenon,
if we thoroughly wash fresh yeast, and mixing a little clear
water with it, let it stand. The formation of the spores here

follows, by a sufficient supply of water, at the cost of the organic substance, which has assimilated during the fermentation; we must seek it in the yeast which is used technically, when, after its fermentation is complete, it is laid aside clear and wet.

The spores begin, when they are brought into a suitable liquid, to sprout like the vegetating cells, in order to produce new repeated generations of the latter. No other forms of development are known, for the fungus which is found in yeast. This produces as the last or first member of its formation, *utricles, Asci,* which are exactly similar to the organs of the same name in other *Ascomycetes. Asci,* wherever we may find them, constitute the last form of development of the fructifying organs of the *Ascomyceta;* it is on account of these facts, that we ascribe the same morphological signification to the yeast fungus, viz: that it is a very simple *Ascomycete,* without any definite division into Mycelium and Conidia. However different the simplicity of its articulation in the course of its development may appear to that, which we have described respecting the mildew fungi, it does not in any way stand isolated or disarranged in the series of fungi forms. It is much more connected with those by a number of parasitical fungi, which infect plants, of the species of *Taphrina, Exoascus,* single kinds of which represent a complete series of links between the yeast form and the typical *Ascomyceta,* which produces Mycelia.

The form and development of the yeast fungus are very simple. The facts we have mentioned with respect to its final fructification, are quite of recent date, and before they were known, the most various views were adopted respecting the nature of yeast. If we omit those views, which consider the yeast fungus to be a spontaneous production and exclusively to proceed from fermentig fluids, and of which the same can verbally be repeated, which has been said above about the presumed spontaneous origin of mildew in the substrate, all the others are founded on the following facts. First, the resemblance of the yeast fungus to typical fungi in every respect, except in the vegetative form. Secondly, the experience, that many typical fungi show similar buddings to those of the yeast fungus, be it in the formation of their Conidia, or elsewhere; also the *Ascomyceta* during the germination of their spores, particularly the above mentioned *Taphrinae* and others, for example, *Dothidea ribesia Fr.*; very many during the formation of their Conidia, and the *Mucor*, which is not *Ascomyceta*, when it forms gems in liquids. Thirdly, the fact that all those fungi in their typical development produce mycelium filaments with fruit-bearers etc. As an organ of fructification could not be found in yeast, it was said, that it was a barren form of fungi, which produces mycelium and is continually budding, a form which is also found elsewhere; and as it always appeared in a particular fluid medium, it was further said, that the nature of the medium determined its form, therefore, a fungus produces in a fermentable liquid yeast cells, while, on another substrate exposed to the air, it forms mycelia with utricles.

Respecting which kinds of fungi especially represent the typically developed form of yeast, opinions are very different. According to some, it is a particular species of mildew; *Mucor Mucedo* and *racemosus*, *Penicillium glaucum*, *Oidium lactis* or any one of these; according to others, all possible kinds of fungi produce yeast in any fluid favorable to them.

It was sought to prove these opinions experimentally in

two ways: the one, by sowing the spores or Mycelia of the
fungus in question in fermentable solutions, and the other,
by sowing yeast on a substrate which is favourable to mildew.
As frequently in the one case — by no means always —
yeast cells and alcohol fermentations were observed after the
sowing of the seed; in the other, *Mucor, Penicillium etc.*, the
proof was considered as certain. It is clear, that in such an
experiment two sources of error were not to be avoided, even
by the most careful seclusion of the substances to be ex-
perimented on, viz: the unintentional intrusion of yeast-cells
with the intentionally sown fungus spores; and on the other
side the mixture of germinating spores of other species of
fungi with the yeast. Further, the typical yeast-cells are
diffused everywhere, even in places infected by mildew,
and that yeast, when made in large quantities, is quite free
from the germs of the common mildew fungus, is scarcely
plausible from the nature of the place and materials, and
in the purest "yeast" almost without any difficulty we
can often find in the most prodigious quantities spores and
gems of *Mucor, Oidium lactis* and *Penicillium.* The germs
of these fungi must preponderate in their development on a
soil suitable to them, at the cost of the less favoured yeast-
cells, the spores mingled in the yeast and sown in the liquid
spring forth at the expense of the germinating utricles, which
proceed from the spores, and the growth of which has been
kept back. That this is the case, is shown in every attentive
examination, in which the observer must not think to replace
his attention by a glass-bell. Experiments to prove an organic continuity between yeast-cells and typical fungus forms,
which alternately grow out of one another, have scarcely been
made; in that tried by *Berkeley*, which appears to show, that

the *Penicillium glaucum* springs from the yeast-cells, the result is explained by the mixture of single *Penicillium* filaments with the yeast which has been used for the purpose.

Besides those sources of error, of which we have already spoken, in describing the forms of mildew, and on which it is not necessary to make any further observation, there is still one special to yeast. With a particular obstinacy it has always been asserted, that yeast is a variety of the two forms of *Mucor*. The principal argument for it was this, that *Mucor* brought into a suitable fluid produces alcoholic fermentation. That is indubitable, but that does not say, that the form or series of form which we call *Mucor*, must be identical with the form *Saccharomyces*, because it has a similar influence on the substrate.

Those *Mucors* of which we speak, form moreover in fermentable fluid those sprouting gems which resemble yeast. In the budding, these are similar to yeast, as for the rest, it is (see Fig. 7, A and B) impossible to mistake them, and the difference of both has always been recognized. Nevertheless, it is said, that *Mucor* produces alcoholic fermentation, with the formation of gems, therefore its germs are yeast, and yeast is a variety of *Mucor*. The impartial reader will be astonished at this, and perhaps seek the source of error in another place, than in the adduced facts; perhaps, he will be still more inclined to do so, when he hears, that the doubts of an unbiassed criticism on the admissibility of the said inference, are indignantly received by the authors of it.

We will now return to our yeast-fungus, the unrefreshing confused debate over which has reached its definitive conclusion with the discovery of its own kind of fructification and germination, and after observing its development of form, we will cast a glance on its relation to alcoholic fermentation, in which it is found, and for which it is employed. It is now, through the works of *Pasteur*, a decided case, that by these fermentations the yeast fungus is the exciter of fermentation, and that it gives the impetus to that which we call fermentation, through its vegetating or nourishing process. Living yeast-cells, capable of growing and budding, are absolutely necessary to the introduction of fermentation, the death of the yeast, for example, by bringing the liquid to boiling point, immediately puts a stop to the process of fermentation; the dead substance of the yeast is in in itself totally ineffective. A minimum of living cells is sufficient in a suitable liquid to cause fermentation; this spreads itself rapidly over the whole mass, as the yeast-cells bud and increase.

Only solutions of sugar are eapable of fermenting, which like *must* and *wort*, contain with sugar the quantity of azotic and mineral nutritious matter, which is requisite for the yeast fungus as woll as every other species. If we add to such more than a minimum of common yeast, we make it fermentable, because the common yeast always contains in its withered cells and their products of disorganization, a quantity of nutritious matter, which changes into a soluble form the sacchariferous fluid, in favour of the living budding cells.

In every fermentation, the number of yeast cells increases prodigiously, as we can see by a superficial observation and weighing, and analysis show most minutely, how an immense increase in the organic and organized substance of the yeast takes place, at the expense of a part of the dissolved sugar and the before mentioned nutritious matter. The remainder of the sugar, that is about 95 per cent of the quantity employed, is divided into alcohol and carbonic acid, with a small quantity of succinic acid and glycerine — that division which is the principal thing by alcoholic fermentation.

We may assume the same hypothesis, based on perfect analogical causes and observations, for the origin of the fermenting action of yeast, as that which we have explained for the same effect in the mildew fungus. The yeast fungus, of which we have hitherto spoken, is the principal promoter of the alcoholic fermentation, which appears in practical life; especially the greater part of beer and spirit fermentation. That which is distinguished by the name of *barm* and the *yeast* deposited at the bottom of the *tube* or *cask,* are in many cases — not in all — the same fungi, which in a lower temperature remains at the bottom and collects as *underyeast*; by higher temperature it accumulates in the froth on the surface of the fluid and is called *barm*. There is a slight difference in the form of the yeasts, but the one form can be transferred to the other by changing the temperature of the fermentation. There are, besides the yeast fungus, others which resemble it and like it produce budding utricles; but the form is that of a well distinguished yeast fungus or species of *Saccharomyces*. Several of those are like the *S. cerevisiae*, exciters of alcoholic fermentation. Thus, the fungus in dregs of wine, or rather fungi, for several well distinguished forms appear in the lees of wine; and another, which differs greatly from the common yeast, and is used in certain fermentations of beer. Most of these forms require more minute examination, particularly with respect to their quality and quantity as exciters of fermentation.

There are, however, *ferment fungi*, that is, forms which may be said to belong to the *Saccharomyces* kind, and are very like *S. cerevisiae*, which do *not* excite alcoholic fermentation. The *mould* on sour wine and beer consists of such a form, the single cells of which are distinguished from those of yeast by a somewhat smaller size in diameter, and in certain difference in the form of structure. The cells are found immersed in the fluid, and slowly vegetating; if the fluid be exposed to the free current of *oxygenated* air, they increase rapidly on the surface, raise themselves over the level of the liquid, and form together the well known white mould. The spores, as far as relates to the spore-utricles of the mould fungus are not yet known; there are, however, indications to show, that they are formed like those of the yeast fungus.

In consequence of the resemblance between the two forms and their frequent appearance in the same fluid, the idea has often arisen, that the fungus which causes mould and yeast, must be one and the same. It is not, however, the case; in spite of their likeness, both fungi strictly preserve their difference of structure and also their different influence on the substrate: the mould fungus excites putrefaction; it oxydizes alcohol and sugar to carbonic acid and water (forming a small quantity of acetic acid, at least in fluid which contains alcohol); it does *not* excite any alcoholic fermentation in sugary solutions capable of fermenting.

There are many botanical names for the mould fungus: *Hormiscium vini, Mycoderma vini, Mycoderma cerevisiae* are mostly used.

Mycoderma signifies a fungus skin or a slimy skin; the name is chosen from the external appearance of the fungus, which is on the surface of the fluid; it originates from a time, in which less attention was paid to microscopic examinations, than to observations made with the naked eye. The name *Mycoderma* is given to other forms of plants, only because they appear in large masses, which with *Mycoderma vini* have only the superficial resemblance we have just described, but are quite different from it in form and development.

Of these *Mycoderma aceti*, in German *"Essigmutter"* (the mother of vinegar), is one of the most remarkable. It leads us from the yeast fungus to another species of "yeast"-form, which still requires a more minute investigation and which we will call the *Bacteria-* or *Schizomyceta*-species.

In the usual method of preparing vinegar, that is, diluted acetic acid, the so-called vinegar mixture — principally diluted alcohol — is exposed, in a suitable temperature, to the influence of the oxygen in the atmospheric air. The alcohol takes up oxygen and is through this oxydated into acetic acid. On the surface of the fluid a slimy scum appears, which sometimes partly sinks to the bottom of the vessel, is again there renewed and has the name of "Essigmutter" (mother of vinegar). It consists of an innumerable quantity of short rod-like little bodies, scarcely $\frac{1}{1000}$ Mm. broad, which prove to be, by closer inspection, vegetable cells, free from chlorophyll. They increase rapidly through separation; after stretching out to a length, about double that of their diameter, they divide into two parts, in each of which the same process is repeated, and an unlimited number of generations are formed. All the points of separation have the same direction, arranged according to their *genetic* connection, and the successive generations of the same origin form a row or chain, which continually increase the number of their members, on all points, by dividing them in half. These chains can sometimes be actually distinguished, they are interlaced and connected with the slimy skin in great quantities, and held together by an

homogeneous slimy jelly. At other times the same little bodies
are often to be found pushed out of their *genetic* connection,
and grasped together within the jelly, single or united into
chains swimming in the surrounding fluid, either motionless,
or more or less quickly moving, that is, oscillating backwards
and forwards in alternate directions. It is ascertained with
certainty, that the same little bodies can alternately assume
either the chainlike or irregular grouping, or the free moving
condition. Further methods of development of the special organs of fructification of this fungus are as yet not known.

Pasteur has shown us, that this little organism brings
about the oxydation of alcohol to acetic acid in the same manner, as the mould and mildew fungus causes the decomposition of the substances on which it vegetates. The formation
of vinegar ceases, when the mixture is exposed to the action
of the oxygen alone, without the presence of the *mother of
vinegar*; it only takes place, when the latter finds the necessary nourishment for its vegetation in the mixture. If we
sink the *Mycoderma* scum to the bottom of a vessel, the formation of vinegar ceases, till a new scum is formed by the
increase of single little bodies, which rise to the surface. If
all the alcohol be oxydated into acetic acid, this will, by con-
tinual vegetation, be burned to carbonic acid and water. In
the quick process of making vinegar, this organism is active,
as it settles on the threads or wood shavings, over which the
mixture flows.

In the solutions which contain sugar, if the necessary
nutriment be present, particular processes follow (under certain
conditions, which have been long known to chemists, and
therefore will not be discussed here), the principal products
of which are *lactic* and *butyric acid*. The experiments, which

have been principally made by *Pasteur*, have shown, that
the cause of their existence and their causal connection is
parallel to that we have mentioned in the fermentation of alcohol and gall. The same sugary solution which, under the
influence of the yeast fungus, is capable of alcoholic fermentation, produces acetic acid fermentation under certain conditions, the principal of which is, the presence of a certain
living organism of fermentation, which may be called *lactic
acid yeast*. According to its organization and development,
as far as the experiments reach, it is not to be distinctly
distinguished from the *mother of vinegar*; only that the free
moveable conditions are relatively more numerous in the former
than it is possible for them to be in the latter.

The same remarks are appropriate to the *butyric acid*
fermentation and yeast. The little twigs, of which they consist, are according to *Pasteur*'s account, on an average longer
and often much longer than those of the *lactic acid* yeast; a
sharp morphological difference is, however, not to be drawn
here.

Pasteur has found similar and analogous effects in a series of less known processes of fermentation, the biography of
which is still to be settled.

At length, those little bodies follow, which are spread
through decaying organic substances, and are known as and
often called *Bacteria, Vibriones, Zoogloea*. That which has
been said about the form, size and increase, the alternate
connection and dissolution, activity and motionlessness of the
members of *vinegar* and *lactic acid* yeast, can simply be repeated when describing those forms, which bear the name of
Bacteria and *Vibriones*, while the name *Zoogloea* is applied
to those groups, which are held together by jelly. There is

no doubt, that of this organism many different spheres exist; still, for the most common of the small forms, which has been everywhere observed and described, a greater distinction from the *vinegar lactic* and *butyric acid yeast* is not possible to be found.

In the dead substances in which we find the *Bacteria*, they are undoubtedly vital promoters of decomposition and putrefaction, which always accompany them, wherever they appear. It may be supposed, that one and the same species and form can vegetate in mediums of a very different special quality, and excites according to the nature of the medium different kinds of decomposition; therefore, one and the same form can appear in diluted alcohol as *mother-vinegar*; in a suitable sugary solution, from which the air is excluded as lactic and butyric acid, in albuminous substances, as the promoter of decomposition combined with ammoniac and sulphurated hydrogen etc. Founded on *Pasteur's* observations, we can express the probability, that those organisms of fermentation require different and even certain mediums, and these contain the different *kinds*, which cause decomposition, the sharp morphological difference of which could not, till now, be settled, in consequence of their resemblance and small size. The decision is to be expected after further examination.

The last mentioned organisms in the manner in which they generally appear, and their fermenting and oxydating action, resemble the fermenting fungus and can therefore be generally designated as yeast. They differ essentially in the development of their form from that of all other fungi, as the vegetating branched thread like rows of cells with the growth at their points or buddings, which we meet with, as the mycelia of mildew and in the fermenting fungi, is here entirely wanting. Their cells increase by dividing themselves into two parts, which division is always uniformly repeated, and the formation of simple cells follows, which *not only* now form the points, but all the other parts.

It is necessary to separate the forms of *mother of vinegar*, *Bacteria* etc. from the fungi into a system founded on the development of form, and to place them with those which are similar to them (but of which we will not speak here), in a particular group, which has received the name of *Schizomyceta*, and which stands in a similar relation to the fungi, as that of those plants, which are not chlorophyll: distinctly distinguished from them in their form of development, but corresponding to them in the conditions and principal appearances of the nutritive process. The *Schyzomyceta* do not stand isolated in the wide range of taxonomy. Just as in the vegetable kingdom plants, which blossom, but are not chlorophyll, are added to certain species, for example the Orchids, Convolvulus etc. and as the real fungi are the next related in form to the green plants, which are known as *Confervae*, thus the *Schyzomyceta* class themselves, according to the present state of knowledge, to a minor group of chlorophyll plants, which bear the name of *Nostocacea*, and their forms which are stately and easily to be observed, agree very often most minutely with that which we know of the *Schizomyceta*. We can speak of the latter as of small *Nostocacea*, which are without chlorophyll, as we speak of Orchids, Gentians etc. which are without chlorophyll.

We have now arrived at the end of our examination, if our task have been to show, by a series of examples, what the botanists know and do not know of "Mildew and Yeast". If we searched further, it would be very long before we finished for just on the examination of the Schizomyceta, all kinds of extraordinary and often repeated stories are united with it.

It has been attempted from many sides, to class the *Bacteria* as a distinct link in the sphere of development of mildew, as *Mucor* etc., founded on experiments and arguments of which we may say the same as that, which we have declared respecting the experiment to prove *Mucor* as the originator of yeast and *vice versâ*. If we proceed a step farther, in such arguments, we encounter the recently propounded *Micrococcus-theory*, the publication of which by its founder must serve as an excuse for us, when we briefly give an account of it.

It has already been said, that the commoner forms of *Bacteria* are widely spread inhabitants of organic substances; their diminutive size and consequently their facility to be removed to another place, thereto their extraordinary great resistance against all external injuries, makes them more suitable than the common mildew fungus, of which that which is similar, has been said. After that which we have learned above, concerning the rivalry of different forms of fungus (and also of plants), the germs of which have come together in one soil, and on which some thrive in consequence of the quality of the soil and the conditions of vegetation, at the expense of the others, it is clear, that a similar relation must and will subsist between the forms of Schizomyceta, and fungi, and experience has very often confirmed this opinion. After all that has been said about the forms and particularly about the small size of the greater part of Schizomyceta, it is very evident, that it is often very difficult to distinguish them from a very finely granulated substance like that which is formed in the cells of plants as the granules of the Protoplasma, or as they appear as the deposits of the most different kinds found externally on living organisms. The granules of such deposits can form chains or be coupled together; they are of a very equal form and size and therefore bear a striking likeness to the real Schizomyceta. A very minute examination is often very necessary, to distinguish them from one another. I remember, for example, a deposit which appeared in wine, which had become sour, which would have been taken by every one for a Schizomyceton in a partial microscopic examination, whereas, in a chemical examination, it was proved undoubtedly as a deposit of tannic acid peroxyd of iron.

If we sow a fungus under conditions, which are not favorable to its development, it will grow scantily or not at all; in the latter case, its spores or utricles often burst, and their granulated protoplasma is emptied out, and then they mingle themselves with the granules of the granulated deposits of the substrate.

Forms of Schizomyceta, which in such seed scarcely ever fail, are often richly developed at the expense of the fungus. They either keep the field alone or predominate in the beginning, to be overtaken later by the yeast- and mildew-fungus, which has been intentionally or unintentionally sown with them, and which grew slowly at first, but at last, perhaps in consequence of the decomposition of the substrate, overgrows and supplants all the others.

In a careful exemination, we can follow step by step the course of this supplanting and overgrowing of the different forms of fungus and Schizomyceta, like that we have several times mentioned above.

It is therefore easily to be understood, why we can often find in one substrate different forms of organism in alternate frequency, near and after one another, together with the deposits and granules etc., which belong to the substrate. If we bring together all these things in one series, just as they are found near and after one another and call it a series of development; if we observe the reaction and the material quality of the substrate, and call them the causes of the different development of form in the series, and at last, if we speak of the most simple members in the series of *Micrococcus*, which appear as minute granules, we have the receipt for a theory, according to which all possible forms of *Schizomyceta* and fungi are successively formed from the *Micrococcus*, and again, that from all possible fungi the *Micrococcus* is formed, and all this through the stipulated influence of the material properties of the substrate.

As far as such a proceeding can bear criticism, we have given it above in several special cases. It would not be becoming in a serious lecture, to enter any further upon the subject. He who simply believes, we do not wish to attack his good and easy credulity. He who has not quite forgotten how to think, will know how to appreciate it, according to its merit and will perceive from the well proved facts, that the organism which we call mildew, yeast and fungi, display certain characteristics, but in their construction and course of life resemble in the principal things all other members of the vegetable kingdom, and that the many and important phenomena, which we observe in them, have their foundation particularly therein, that they are plants like all others, only smaller than the greater part, and therefore, when examined, require more care and attention.

Printed by **U n g e r B r o t h e r s** (Th. Grimm), Schönebergerstr. 17a. Berlin.

Reprint Publishing